THE NATIONAL MUSEUM OF NATURAL HISTORY

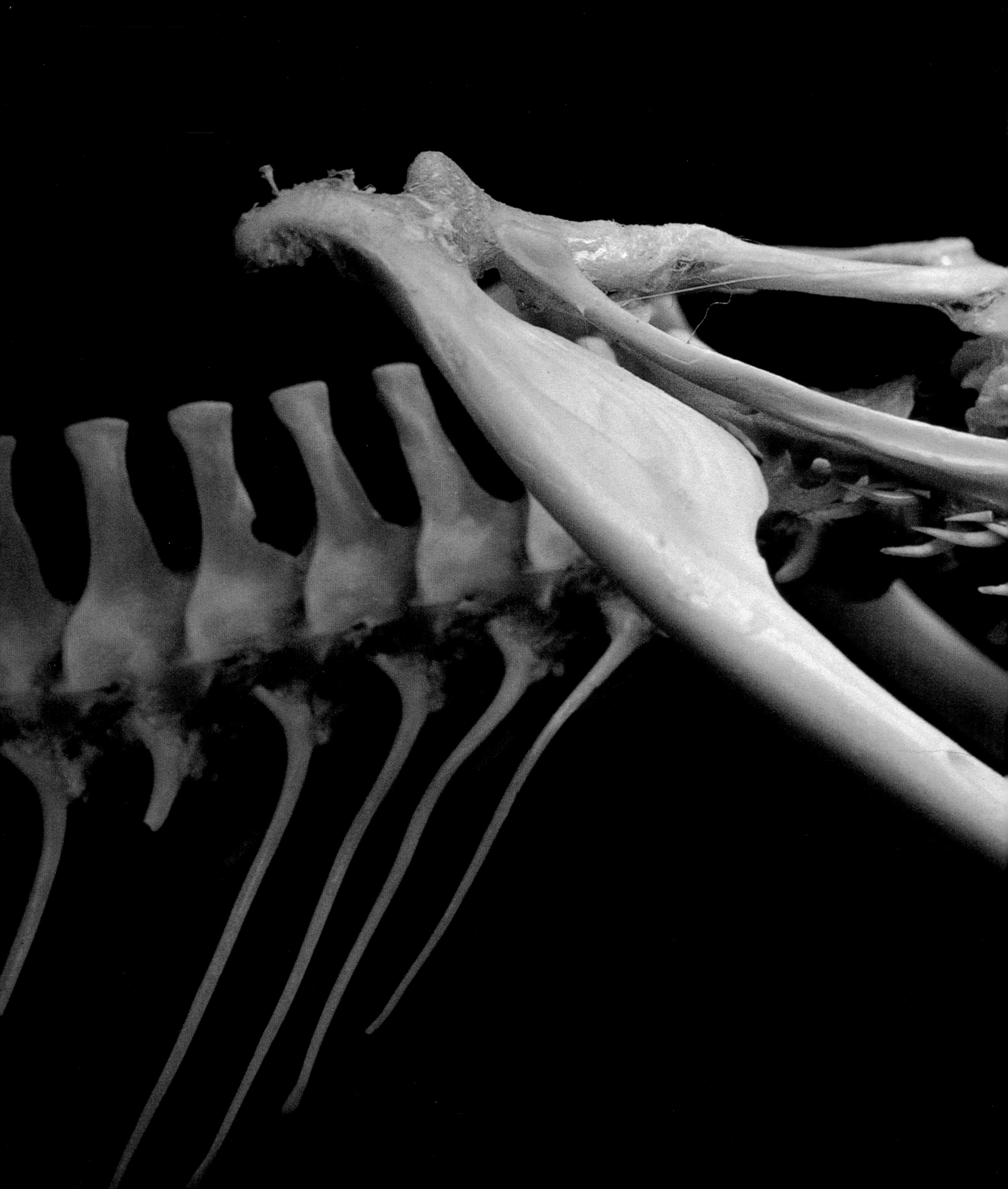

NW COAST
MASKS
Bella Bella
20,570 20,582
20,576-7-8 74,752
97
8893

20576
36649

THE NATIONAL MUSEUM OF NATURAL HISTORY

By Philip Kopper

Photographs by Kjell Sandved and Chip Clark

Harry N. Abrams, Inc., Publishers, New York

In association with the Smithsonian Institution

When silicon and oxygen atoms combine, as in this natural amethyst, their atomic geometry dictates the growth of hexagonal crystals.

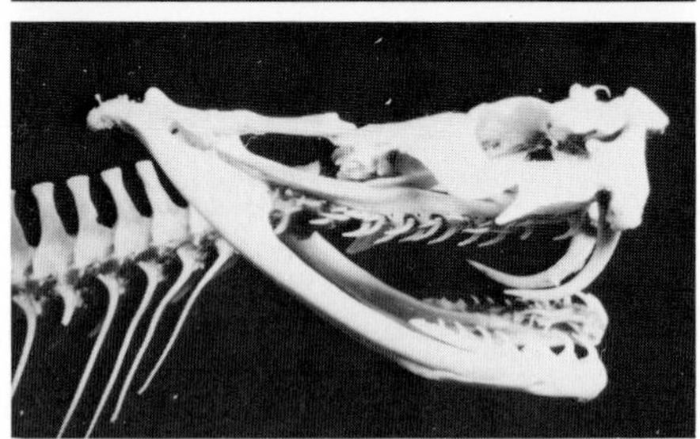
From tropical Africa, the skeletal perfection of a Gaboon viper. The enlarged fangs move to an upright position during a strike and fold inward as the mouth closes.

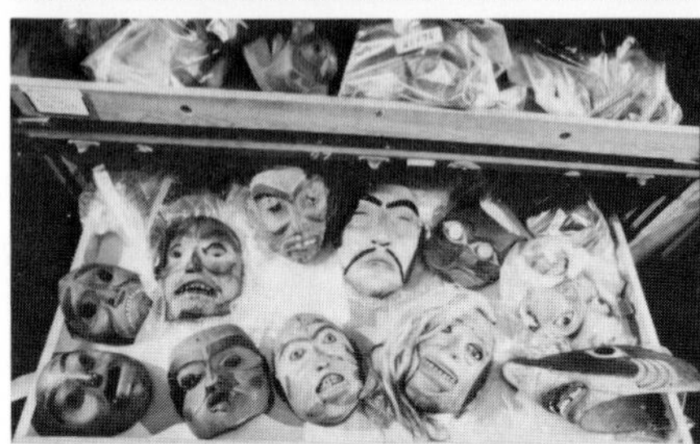
Fine art from the Pacific Northwest —Indian masks in the anthropology research collections.

From the rapidly evolving world of orchids, a species whose lips quake in the slightest breeze to attract pollinating insects.

From the cedar swamps of southern Appalachia via a taxidermist's shop and artist's palette, a diorama of whitetail deer.

From the forests and meadows of Southern Asia, the spectacular butterfly Armandia lidderdalii.

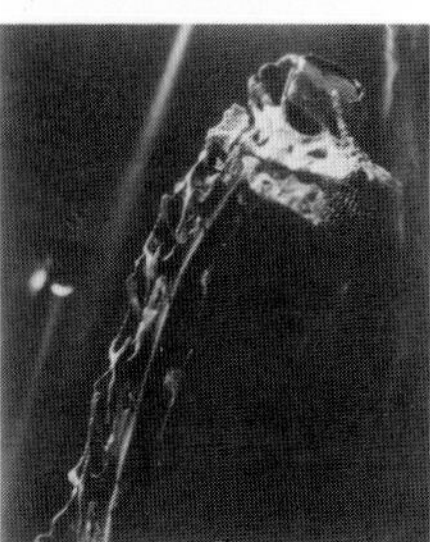
From Utah 145 million years ago, Diplodocus longus, *another of the 60-odd million objects to admire, ponder or study in the Museum of Natural History.*

FOR MARY, MY WIFE

Editor: Edith M. Pavese
Designer: Patrick Cunningham

First Edition

Library of Congress Cataloging in Publication Data

Kopper, Philip.
The National Museum of Natural History.

Bibliography: p.
Includes index.
1. National Museum of Natural History. I. Title.
QH70.U62W274 508 82-1715
ISBN 0-8109-1359-3 AACR2

Published in 1982 by
Harry N. Abrams, Incorporated, New York

Printed and bound in Japan

ACKNOWLEDGMENTS

One man could no more begin exploring this Museum solo than visit all earth's habitats alone. I was blessed that Herman J. Viola, director of the National Anthropological Archives, was my guide at the start, then stayed ready with useful advice and sage insights. When it came time to confirm 400 captions with nearly as many specialists, he pursued that Sisyphean labor with energy and patience. Advisor and friend, he has my sustained gratitude.

I also happily thank the ad hoc review committee whose diversely disciplined members read the text and caught gaffes hidden in my foolscap. Any textual errors are mine, of course; that there are not more is because these widely aware scientists generously contributed such constructive counsel: paleozoologist (and former director) Porter M. Kier, entomologist Paul J. Spangler, Vertebrate Zoology chairman George R. Zug, botanist Stanwyn G. Shetler, Invertebrate Zoology chairman Clyde F. E. Roper, mineralogist Brian Mason.

The remaining department chairmen and heads of various offices and collections opened their doors to collectively give me the run of the Museum. For the knowledge each of them shared, I am grateful to Douglas K. Ubelaker (Anthropology), Daniel E. Appleman (Mineral Sciences), Martin A. Buzas (Paleobiology), Dieter C. Wasshausen (Botany), John S. White (Gems), Eugene F. Behlen (Exhibits), Phillip J. Angle (vertebrates), Gary F. Hevel (insects), Frederick C. Collier (fossils), Roy S. Clarke Jr. (meteorites), U. Vincent Wilcox (artifacts), Carolyn L. Rose (Anthropology Conservation Lab).

Others put aside their work many times to offer intensive courses in esoteric specialties, answers to key questions and friendly support. I am pleased to acknowledge varied and substantial debts to them all: physical anthropologist J. Lawrence Angel, corals authority Frederick M. Bayer, microlepidopterist Donald R. Davis, archivist William A. Deiss, medical entomologist Michael E. Faran, archaeologist William W. Fitzhugh, mammalogist Charles O. Handley Jr., paleomammalogist Jessica A. Harrison, herpetologist W. Ronald Heyer, paleobotanist Leo J. Hickey, USDA entomologist Ronald W. Hodges, invertebratologist W. Duane Hope, the late entomologist Paul D. Hurd Jr., worms authority Meredith L. Jones, librarian Jack K. Marquardt, marine mammalogist James G. Mead, anthropologist Betty C. Meggers, visiting entomologist Eugene Munro, physical anthropologist Donald J. Ortner, marine biologist David L. Pawson, botanist Robert W. Read, paleontology specialist Raymond T. Rye III, archaeologist Dennis J. Stanford, mammalogist Richard W. Thorington Jr., geologist Kenneth M. Towe, illustrator George L. Venable, Insect Zookeeper Kay Weisberg, editor Sue Willis. This group also includes valued advisors on my earlier natural science book *The Wild Edge:* crustacean curator Raymond B. Manning, sedimentologist Jack W. Pierce, mollusk authority Joseph Rosewater, librarian Ruth Schallert, NMFS taxonomist Austin B. Williams, ornithologists George E. Watson and Richard S. Zusi.

From the Secretary in his Castle to the keeper of the beetles, many individuals helped on specific occasions or simply through their abiding interest: Walter H. Adey, Edward S. Ayensu, Harold H. Banks Jr., Marguerite Brigida, Fenner Chace, David Challinor, J. F. Gates Clarke, Stephanie Damadio, Paul Desautels, Albert Diaz, Maureen E. Downey, Nigel Elmore, Terry L. Erwin, Kristian Fauchald, Candy Feller, Richard S. Fiske, Paula Fleming, Kurt A. I. Frederiksson, T. Gary Gautier, Gordon D. Gibson, Leroy Glenn Jr., Francis M. Greenwell, Thomas R. Harney, Elaine R. Hodges, Nicholas Hotton III, Francis M. Hueber, Eugene Jarosewich, Douglas K. John, Adrienne L. Kaeppler, Catherine J. Kerby, Arnold D. Lewis, Felix C. Lowe, Joan Madden, Laura L. McKie, Joan Miles, Richard Molinaroli, Joseph A. Nelen, James N. Norris, Storrs L. Olson, Elaine Peret, Kristine Ramaekers, S. Dillon Ripley, Barbara Stuckenrath, Joseph M. Shealy, Bruce Smith, Thomas R. Sodestrom, Victor G. Springer, T. Dale Stewart, William C. Sturtevant, Paul Taylor, Vyrtis Thomas, Gus W. Van Beek, Catharine Valentour, Suzanne Villemarette, William W. Warner, Deborah Wood.

Next I applaud photographer Chip Clark for his energy, care and skillful ingenuity in taking so many of the striking original photographs in this book. Senior photographer Kjell Sandved also made substantial contributions through new work and especially by making available his legendary slide files.

Finally, I am grateful to the people at Harry N. Abrams, Inc., who first made this book possible and then a reality. My special thanks go to editor Edith Pavese and designer Patrick Cunningham for dedicating their skills to the volume which reveals the dimensions of their talents.

—P.K.

CONTENTS

Opposite: *This marine snail's shell from Taiwan, called the Miraculous Thatcheria, appears on display in the "Splendors of Nature" exhibit.*

FOREWORD

Curiosity is a trait that man shares with many of his fellow creatures, but none of the others are quite so assiduous as man in its practice. Curiosity leads to interest, interest leads to study, and study leads to knowledge. Over the centuries, curiosity about the natural world has led mankind toward an incredible accumulation of knowledge, the fruit of curiosities kindled in generations of naturalists and scientists.

At the National Museum of Natural History, this knowledge is manifested in three important ways. There is knowledge that is derived from the research carried out by the Museum's staff of scientists in biology, anthropology, and the earth sciences. There is knowledge inherent in our collections, which consist of more than 60 million specimens, held as a form of trust for present and future generations of scientists. And there is knowledge manifested in the Museum's exhibits and education programs, which serve as the means by which we speak to the more than 5 million people who visit our public spaces each year.

From the earliest "cabinets of curiosities," assembled centuries ago by a few of the leisure class in Europe, to today's collections of the National Museum of Natural History, there is an enormous jump in scope and reach, and yet it is the basic stimulus of curiosity that fostered the accumulation of both kinds of collections. An important difference between then and now is the attention paid to these curiosities. Then, they were just peculiar things—a stone with a shell in it, a single twisted horn, or an enormous tooth. Today, after 300 years of work on nature's puzzle, the patterns of nature are largely understood, and the place of each object in these patterns can be comprehended at a much higher level of understanding. The scope of this enterprise, begun with a few curiosities, has now extended to every kind of creature, both great and small. Not only do we work on elephants and orchids and feather headdresses, but also on banana slugs, and seaweed, and kitchen middens. Our collections even extend to the intangible—to the vocabularies of nearly forgotten languages and to the vast store of data on eruptions from the world's 600 active volcanoes.

This book presents an all-important dynamic, the constant feedback from research and exhibits to collections, and vice versa, that best defines

This rattle, once owned by a shaman of the Tsimshian Indians of British Columbia, was described a century ago by its collector as representing the wind spirit T'kul and progeny, and indicating "how frogs and toads come with the rain."

what we are all about. Author Philip Kopper has captured this spirit perfectly. By treating our Museum as a series of museums (a museum of collections, of conservation, of science, and of exhibits), he has defined various aspects of our operation that, in most cases, slice obliquely across the discipline-related makeup of our seven scientific departments. But look carefully at each of his "museums," and you will recognize the all-important threefold elements—collections and their management, the pursuit of scholarly research, and outreach via exhibits and educational programs.

In the course of his work, Philip Kopper literally became a member of our Museum's family. Not only did he turn up at exhibit hall openings, Museum and departmental Christmas parties, and other inhouse assemblies, but he could also be found lurking in our attics (yes, we have more than one!) and wandering through our cellar. He became intimately familiar with our collections, with our staff, and with our public spaces. In the process, he captured the elusive spirit of what the Museum is all about.

Having begun this Foreword by reflecting on our heritage of past traditions, it is appropriate to close with a view toward the future. The animals and plants with which we share our fragile world live in their environments within limits of tolerance, and in close association with other living creatures. These limits and associations cannot be perturbed without jeopardizing our fellow travelers on earth. We have striven to become masters of the globe and, over our history, have exploited its resources for our gain—following, perhaps unconsciously, the Biblical mandate to multiply and subdue the earth. Our industrial requirements, our agricultural practices, and the very waste from our enormous numbers, however, are placing unprecedented stresses on the natural world. Numerous species of plants and animals, the result of evolutionary processes operating during millions of years of earth history, are being irrevocably lost each year as a result of these stresses. While no single institution seems capable of halting these changes or altering harmful practices, museums such as the National Museum of Natural History can at least continue, and even accelerate, their traditional role of gathering in representatives of the life and cultures with which we share the world, before they disappear.

The Museum of Natural History's "cabinet of curiosities" has grown through the flourishing of our curiosity, and this curiosity must continue, so that we can preserve the vestiges of what is being destroyed, understand the complex interrelationships of what will remain, and inform the public as to what is happening in the natural world. By pushing forward with our threefold mission of collections management, research, and outreach, we shall achieve these goals. This is an important mission, and the pages of this magnificent book reveal some of the ways in which the Museum is carrying it out. Read on!

Richard S. Fiske
Director,
The National Museum of Natural History

PROLOGUE

Welcome to all the worlds of earth.

Welcome to a sanctuary of serene sciences and bizarre objects, a repository of the extraterrestrial and of the subterranean, a treasure trove of exotic human artifacts and pickled biological specimens. Mummies and monsters abide here with dinosaurs, the Hope Diamond, the oldest rock yet found, the earliest fossil known to man, the latest unidentified corpse known to the FBI. There's the Winchester rifle that Chief Sitting Bull brandished at Little Big Horn. Here's a deepsea worm that grows five feet long and thick as a garden hose though it never eats. There's a flying reptile with a 40-foot wingspan. There are rocks drilled clear through by lightning and reed baskets woven so tight they hold water and parkas made of fish gut. A plaster bust of Anna's King of Siam stands in one locked storeroom and the hide of General Pershing's horse lies in a drawer. Two rooms big as roller rinks hold the most complete collection of plants in this hemisphere. Elsewhere a vault the size of a powder room holds paste "jewels" that an Arab prince foisted on a Vice President. Genuine shams abound too: live bugs that move like leaves wafting in the wind; iridescent butterflies whose wings have no color but refract light like prisms; mastodons and microbes long since transformed by the alchemy of nature.

Welcome to a world of the infinite and the infinitesimal: the suction cups on a fly's foot enlarged to demitasse size; the Old Woman and Lost City meteorites which brought molecularly frozen proof of this tiny planet's explosive birth some 4,600,000,000 years ago; "Baby Doll" (a.k.a. *Thrinaxodon liorhinus*), an archaic oddity of a reptile whose descendents survive only as mammals. Welcome to a place of entertainment and mystic drama: a theater with live actors; boxes of Indian children's totem toys; cases of Hindu dancers' masks showing the great god Shiva in his many forms. Welcome to a place of peculiar work: where acids wash ancient trilobites from their tombs of stone; where technicians wielding dentists' drills release herbivores the size of hippos from solid rock millimeter by millimeter; where a doctor of anthropology makes bone tools to dismember a dead elephant called Ginsberg; where tireless beetles clean skeletons with neurosurgical precision. In one basement cavern human volunteers have spent the past decade piecing together shards of Palestine pots

Symbol of the Museum and its ceremonial sentry since 1959, the African elephant charges in perpetuity beneath the Rotunda. One of the largest of its kind ever seen by men, this mighty bull stood 15 feet tall at the shoulder, weighed 8 tons and may have lived for nearly a century.

This Museum's uncounted marvels represent many sciences, come from every nation of the world and span all earthly time. Among them: A butterfly's magnified wing scales which contain no pigments but like prisms refract white light into component colors; a rifle carried by Sitting Bull, the Sioux chief who vainly tried to hold onto his traditional way of life; the decorated lid of the coffin which held the mummy of Tenet-Khonsu, an Egyptian high priestess to the god Amon-Ra about 1000 B.C.

broken before Christ or Moses walked the local byways. In the upstairs laboratories specialists spend their lives burrowing through arcane data to explain the phylogeny of sea fans, the life cycles of water beetles, the mechanics of frogs' legs, the legacies of giant ferns, the habits of mankind and countless other esoterica. Meanwhile quiet corps of technicians preserve specimens of almost every plant and animal—extant or extinct—ever seen on earth.

Welcome to a building containing 100 million cubic feet of space that 5 million visitors entered last year: jamborees of Boy Scouts to decipher a film on genetic coding; bevies of Pennsylvania Dutch farm wives to admire the handiwork of vanished Polynesian women; foreign delegations to encounter strange beasts that will never again range the Great Plains; schools of children to whisper at parrotfish browsing the first coral reef to live and grow apart from the sea. Adolescents, bachelors, businesswomen, lovers, paleophiles, priests—people of every imaginable sort—ponder the equally diverse collections on display. Yet they see only a fraction of this place and the objects it contains because barely a fifth of its floors accommodates the passing throng. The rest is dedicated to some 1,200 identifiable rooms and spaces: laboratories, libraries, workshops, study halls, classrooms, darkrooms, lead-walled radiation chambers, a mail room that handles 7 million items a year, attics packed with stuffed animals, a fur vault hung with the pelts of Teddy Roosevelt's hunting trophies, a gem room which no stranger may enter alone, autopsy cubicles, labyrinths of hallways stacked to the ceilings with carefully boxed and labeled human skeletons, a freezer big enough to preserve the world's one remaining "dragon," locker rooms; paint, machine, electrical and cabinet shops; "ranges" big as basketball courts filled with reptiles in jars of alcohol, 4 million specimens of plants pressed on sheets of paper, 26 million insects pinioned on pins in little boxes, data banks, a greenhouse, and nearly 400,000 square feet of storage. That's enough floor space to accommodate 200 homes or 50,000 hall closets, yet room that often reaches 40 feet up and contains dusty relics of the human or animal past from stone floor to vaulted roof. (No one has counted the objects convincingly. Even the inventory computer fails at that task because many items—a quiver of

COLLECTION OF FISHES
9-73

Pages 28–9: *Recovered from a California tar pit after burial 14,000 years ago, a saber-toothed cat bares its fangs again.*

Pages 30–1: *Carefully bottled and shelved in a storage range, many thousands of specimens represent nearly every species of fish known to science.*

Opposite: *The Spanish Inquisition necklace, made more than three centuries ago of emeralds from the New World and diamonds from India, is one of many Museum treasures of inestimable—or unreportable—worth.*

spears or a jug of sea cucumbers—constitute a collection in themselves.)

Welcome to the place where the many natures of earth have been catalogued, examined and revealed, where they continue to be explained through seven disciplinary departments: Anthropology, Botany, Entomology, Invertebrate Zoology, Mineral Sciences, Paleobiology and Vertebrate Zoology. Among them, these sciences reach into every realm of the natural world: its people and plants, its insects and spineless animals, its fossils of obsolete organisms, the physical earth itself, the cosmos which created that and all the animals inhabiting this planet including the genus *Homo* in our many guises.

Welcome to a Mecca for scientists from every realm and present nation who comb these collections for old clues to new knowledge. Welcome to a repository in perpetuity that continues to grow as scientific packrats and philanthropists endow it with the fruits of their private passions and eccentricities—their collections of everything from priceless gemstones to pachyderm skins. Welcome to a monument that earns the affection and awe of curious people in every walk of life because it entertains and educates us all. Welcome to a temple of earth's physical, biological and human experience, a launching pad for man's future explorations —whether via electron microscope or interplanetary spacecraft—into every natural realm. Welcome to a sanctuary where relics of all earth's past, as all the sciences slowly discover them, safely stay for the eventual edification of mankind. Welcome to this repository of the exotic and the base, the legendary and the physical, the petrified and the alive.

Welcome to the National Museum of Natural History, a microcosm of earth, the Smithsonian Institution's protean pantheon to our world's myriad and magnificent mysteries.

THE EVOLUTION OF THESE MUSEUMS

The National Museum of Natural History was never really founded. It evolved. Not conceived in the mind of one person, never officially chartered by name, it was nurtured by a parade of people. It appeared as a child of its time, a product of manifest destiny: spontaneous, omni-directional, unplanned and insatiable. Like the eccentric Briton who endowed its parent organization, the Museum was born a benign bastard. Like its first collections supervisor, it recognized no limits to its concerns, curiosity or energy. Its growth, its very being have always been organic. The story of its maturation combines financial serendipity, mindless dedication, quixotic compulsion, catholic cooperation, bizarre genius and blind luck. The servant (and sometimes master) of the scientists who inhabited it, the Museum was long a pioneering force in the natural sciences. Inevitably and purposefully it became more conservative in both the philosophical sense of the word and the elementary sense; while many of its people still march in the vanguards of the sciences, the majority of its energies are devoted to conserving the palpable sources of knowledge, i.e., objects. Once driven by inspired whim, this institution within the Institution has become the stately steward of natural history. At once stodgy and vital, the Museum of Natural History is productively organic and purposefully unique—perhaps because nobody planned it that way. In fact it was hardly planned at all. It just grew like one of the insects or crustacea its taxonomists would study: by fits, starts and metamorphosis.

This grand turn-of-the-century neo-classical design, proposed for the new National Museum, was deemed too spectacular. Lest it compete with the U.S. Capitol nearby, Washington architects Hornblower & Marshall scaled down their plan in favor of the more modest building.

Though the Museum had no discrete birth, its secret conception occurred in London on October 23, 1826, when a malcontent aristocrat composed his last will and testament. Nine years later the document came to the attention of Aaron Vail, the United States chargé d'affaires in Britain who wrote a copiously circumlocutory letter to his superior in Washington, the Secretary of State.

> Sir: The papers which I have the honor herewith to communicate to you will acquaint you with the particulars of a Bequest of property to a large amount left to the United States by a Mr. James Smithson. . . . The caption of the Will is in language which might induce a belief that

James Smithson, English bachelor of gentle if clouded birth, sat for H. Johns, who painted this miniature. Short of stature and long on eccentricity, the gentleman scientist bequeathed his considerable fortune to establish in a land he'd never seen a namesake institution "for the increase & diffusion of knowledge among men," which in turn begat the National Museum of Natural History. The oddities that tinged his life did not abate with his 1829 death in Genoa, whence his remains were transported to Washington 75 years later. When the sealed casket was opened with a blowtorch the interior began to smolder; rather than use chemical extinguishers, attendants quenched the flames with water brought from a drinking fountain—in their mouths. Subsequent examination revealed that Smithson stood 5 foot 4 and possessed an extra vertebra.

> the Testator labored under some degree of mental aberration at the time it was made, tho I understand that its allegations are not destitute of probability, at least.

Whatever Vail meant exactly by that last sentence, it was bracketed by an early reader and deleted from later copies, especially from official histories of the eventual institution. Perhaps Secretary of State John Forsythe himself decided not to let the young nation look a gift horse in the mouth too closely. America would have enough trouble accepting the legacy without examining the donor's state of mind.

The late Mr. Smithson, a scientist of considerable talent and reputation, enjoyed the intimacy of Britain's notable natural philosophers. He was an eccentric, understandably perhaps, who possessed inherited wealth which allowed him to coddle a constructive mania. After taking a degree at Pembroke College, Oxford, and being nominated to membership in the Royal Society at the age of 22 by no less a light than Henry Cavendish, he devoted his adult life to the infant science of chemistry. He rambled Scotland and the Continent in its pursuit, going from place to place with a portable laboratory and assaying everything from the saline constituents of a lady's tear to the unidentified minerals he stumbled upon. One of them, a zinc carbonate which he analyzed for the first time, was named *smithsonite* in his perpetual honor. Unencumbered by academic responsibilities, he wrote scientific papers which appeared from time to time in the most learned journals. Lacking a family, he dedicated his energies to work and for diversion gambled when the spirit moved him.

Yet he remained a deeply dissatisfied man. "The best blood of England flows in my veins; on my father's side I am a Northumberland, on my mother's I am related to Kings, but this avails me not," he wrote. Born under the bar sinister, he was barred from Court, elite society and the established professions. Illegitimate son of the first Duke of Northumberland and a wealthy widow of Bath, he vowed "My name shall live in the memory of man when the titles of the Northumberlands . . . are extinct and forgotten." The vengeful motive behind that boast may have been what inspired his largesse toward a country he had never visited.

Precisely what Smithson had in mind when he wrote the will remains anybody's guess. He left a small annuity to a servant and the rest of his estate to a nephew with a proviso for its disposal if that bachelor died without heirs (pointedly either "legitimate or illegitimate" ones). The ultimate charge was simple enough:

> In the case of the death of my said Nephew without leaving a child or children. . . . I then bequeath the whole of my property . . . to the United States of America, to found at Washington, under the name of Smithsonian Institution, an Establishment for the increase & diffusion of knowledge among men.

A university? A learned society? An eleemosynary foundation, library or laboratory? Any or none of the above so far as his testament went. An Establishment.

This vaguery was hardly addressed for ten years. First the Washington establishment had to debate the gift itself. President Andrew Jackson, who preferred to deal with Britons at the end of a rope when he dealt with them at all, informed Congress he had no authority to accept the bequest. Via a letter now displayed in the Secretary's office, he passed the buck to Congress. Senator John C. Calhoun deemed it "beneath the dignity of the United States to receive presents of this kind from anyone." Another South Carolinian in Congress considered the gift a cheap shot at immortal respectability; to accept it would be an open invitation to "every whipper-snapper vagabond" who wished to distinguish himself.

Former President John Quincy Adams, then a solon in the House of Representatives, thought otherwise. He persuaded his colleagues to at least receive the gift and Jackson signed the authorizing bill. Richard Rush, a Philadelphia lawyer and son of a signer of the Declaration of Independence, was appointed to wrestle the bequest through the British Court of Chancery. A better man couldn't have been found for the job. He had served as Attorney General, Treasury Secretary and Secretary of State. As Minister to the Court of St. James in 1818, he was credited with soothing British tempers and avoiding war when then Major General

Andrew Jackson hanged two royal subjects for trafficking with intractable Seminoles. Working a minor miracle, Rush got the will probated ahead of 800 pending cases and brought home £107,960 (in gold sovereigns) 8s. 7d. along with the deceased's laboratory gear, mineral collections and personal linen. (His remains arrived many years later.) Back in Philadelphia the fortune was melted down and minted into coin of the republic worth $508,318.46 on July 20, 1838.

That amount, deposited in the Treasury, equalled all the costs of running Congress, defraying House and Senate expenses and paying House members' salaries. A substantial sum, it was soon put to ill use, converted to bonds which were never redeemed. The original money vanished as a discrete fund, but Congress would make it up with interest to the infant Smithsonian when the time was ripe.

Barely three years passed between Vail's letter and Rush's return with the money. Congress then alternately dawdled and harangued for eight years more before deciding what to do. It was Adams again who carried the day by lobbying tirelessly for a well-defined cause, even making a speech in its defense the night a grandchild lay on her deathbed. Opposing various other proposals—such as to rescue a floundering university, build an astronomical observatory or spend the trust on training teachers—Adams wanted to follow Smithson's instructions as closely as possible. He finally got his way in Congress, and President James K. Polk signed the Act of August 10, 1846. It established the Smithsonian Institution under a Board of Regents and endowed it with some peculiar powers. But the document was so unspecific in its particulars that the organization was nearly a dummy corporation, albeit one "for the increase and diffusion of knowledge among men." What the outfit would actually do, and how, was left largely for the Regents to decide. What they would do it *in* was somewhat clearer because Congress mandated that: The Regents

> shall cause to be erected a suitable building, of plain and durable materials and structure, without unnecessary ornament, and of sufficient size, and with suitable rooms, or halls, for the reception and arrangement, upon a liberal scale, of objects of natural history,

> including a geological and mineralogical cabinet; also a chemical laboratory, a library, a gallery of art and the necessary lecture rooms. . . . As suitable arrangement can be made for their reception, all objects of art and of foreign and curious research, and all objects of natural history, plants, and geological and mineralogical specimens belonging, or hereafter to belong, to the United States, which may be in the city of Washington, in whosoever custody the same may be, shall be delivered to such persons as may be authorized by the Board of Regents to receive them, and shall be arranged in such order, and so classed, as best facilitate the examination and study of them.

Clearly the goal was a lyceum of sorts where scholarly lectures could be delivered. It would contain an art gallery and library on equal footing with the Library of Congress. It was also to be a repository of natural history specimens arranged for study by scholars, possibly a "museum." But the Congressional authors pussyfooted and only used that word in passing, as when they declared "the said Secretary of the Board of Regents shall . . . discharge the duties of librarian and keeper of the museum."

The Regents decided their chief executive officer should be a man of "eminent scientific and general requirements." For their first Secretary they chose Joseph Henry, one of the nation's most distinguished scientists, the first American since Benjamin Franklin to be regarded here and abroad as an original investigator of importance in physical science. The job description called for a man capable of "advancing science and promoting letters by original research and effort, well qualified to act as a respected channel of communication between the Institution and scientific and literary individuals and societies in this and foreign countries." Then a professor at the College of New Jersey in Princeton, Henry certainly met those particulars. He was the first experimenter to make a telegraph work (between two campus buildings), though Samuel F. B. Morse later patented a similar device and contrived the code which made it a practical tool. His electromagnetic experiments were epochal and he applied his interest in meteorology to set up a national network of volunteer observers who used Morse's telegraphs to send him daily weather reports. As Secretary,

Left: *Before the new Museum's construction, natural history specimens were displayed in the Arts & Industries Building.*

Right: *Joseph Henry, one of the premier scientists of his day and the Smithsonian's first Secretary, stood for his Napoleonic portrait taken by photographer Matthew Brady.*

Henry launched an international system for the exchange of scientific literature which survives today—as does his meteorological web in the National Weather Service. He also began the Smithsonian's extraordinary publications program and its keystone serial "Smithsonian Contributions to Knowledge." Investigator and visionary, he possessed political abilities that often allowed him to get his way with Congress. A Renaissance polymath of sorts, he was not, however, a "museums man."

Henry was loathe for the Institution to conserve and curate things of almost any sort except those required for experimental work. "All unnecessary expenditure on local objects would be a perversion of the [Smithson] trust," he once told Congress. One of his first assistants was Charles Coffin Jewett, a gifted bibliophile hired to supervise the Smithsonian's library. When he lobbied to spend the lion's share of the budget on book acquisitions, Henry fired him. Jewett was one of the great librarians of his day and his dismissal ended the Smithsonian's mandated role to build a great library. Though the action ran against the letter and spirit of the charter, the Secretary persuaded Congress to look the other way. (The bulk of the Institution's books and all its brief copyright responsibility went to the Library of Congress in 1866.) Some failure or oversight also occurred in the building of the Castle on the Mall, an edifice costing $250,000 that hardly conformed to the Congressional vision of a sensible structure "without unnecessary ornament." However charming it looks to us today, James Renwick's fabulous Romanesque design seemed more suitable for a twelfth-century monastery.

In 1849, before it was finished, Henry moved his family and equipment into one of the eight towers and continued the work he had already begun: "To Increase Knowledge. . . . To stimulate men of talent to make original researches by offering suitable rewards for memoirs containing new truths. . . . To appropriate annually a portion of the income for particular researches"; then "To Diffuse Knowledge" through "a series of periodical reports" and "occasionally separate treatises." Frustrated by administrative chores, in 1850 Henry had hired a remarkable assistant to handle those tasks and look after the Institution's possessions. Spencer Fullerton Baird came with John James Audubon's recommendation and a penchant for collecting things. The consummate packrat, he brought with him two boxcars containing 3,600 birds, 500 jars of reptiles and fishes, 600 skeletons, sundry mammals and some fossils—all of which he donated to the Smithsonian as the germ of its collections. A congenital collector and a kind of whirlwind genius, he worked harmoniously with Henry whatever misgivings the latter had about collections and museum functions. Succeeding Henry in 1878, it was Baird who started the Smithsonian on its way to become famously a community of museums.

The first steps in that direction, however, were taken during Henry's administration when the Secretary contrived a ploy that went awry. It involved a rival organization, the National Institution for the Promotion of Science and the Useful Arts (more economically the National Institute after 1842). Joel R. Poinsett, its guiding light and the man for whom the poinsettia is named, had designs on the Smithson bequest before he fell out of Congressional favor. Lacking a building of its own, the Institute stored its "National Cabinet of Curiosities" in the Patent Office until these collections devolved to the Commissioner of Patents. By the mid-1850s that worthy was agitating to get the burgeoning material out from under his roof. Meanwhile, Henry had found the Castle not to his liking and conceived a mutually advantageous shuffle. He proposed that Congress buy the new building, build display cases, install the National Institute material and run it as a public museum. The Smithsonian, he offered somewhat plaintively, would continue to occupy only "a few rooms" for its work of stimulating research and distributing scientific literature.

In its wisdom Congress bought all these suggestions but one. The material was shipped to the Castle in 1858 and installed in newly bought cases. But the Smithsonian was still in charge of the entire place. It had acquired these collections and the unavoidable responsibility of exhibiting them to the public. To ease the burden, however, Congress provided public money to support the museum functions it foisted on Henry.

So Henry found himself in possession of a museum and gamely proceeded to follow instructions—with Baird's considerable assistance. When the material went on display a sign appeared in the hall bearing a caption that hadn't been officially authorized so far as anybody can tell. It read "National Museum of the United States." Archivists have been unable to discover whose brainchild that was, but the name stuck and within two decades had been institutionalized in Congressional documents.

The first structure designed as a museum, the multicolored brick Arts and Industries Building beside the Castle, was raised to display the best of the things exhibited at the Philadelphia Centennial Exposition of 1876. Baird supervised its design, demanding that it be as cheap and commodious as possible. It had to be. As archivist William A. Deiss has written, "During his 28 years as Assistant Secretary the number of specimens in the collection increased from about 6,000 to several hundred thousand with an additional 400,000 specimens distributed to other museums and individuals. By the time of Baird's death in 1887 the holdings of the museum totalled over two and a half million specimens."

Baird must have been the Henry Ford of American museums—not the man who invented the prototype, but the one who made it work as a vehicle for collection, preservation, research, education and exhibition. He himself had caught the collecting bug as a teenager who'd often cover 40 miles a day with pack and gun around Carlisle, Pennsylvania. By the time he was seventeen he'd corresponded with Audubon and described a hitherto unknown flycatcher. In his nineteenth year he wore out his boots three times, walking 2,000 miles and shooting 128 species of birds. After a brief stint as a medical student in New York and a vain stab at a National Institute job, he put in several years at Dickinson College teaching natural history, tending a museum and serving as librarian.

When he came to Washington his new duties curtailed his freedom to collect in the field. Instead he found ways to do it vicariously by recruiting a huge corps of volunteers around America. His connections in the scientific community, already impressive, became vast to reach a network of farflung amateurs as well as august academics. His energy remained boundless and he happened to come along at a fortuitous time. The federal government was mounting numerous exploratory expeditions: to survey the nation's transcontinental boundaries, to map railroad and wagon routes, simply to chart the West. The archivist writes that Baird "found naturalists to accompany the expeditions, encouraged military personnel to assist them, and even encouraged the soldiers themselves to collect in their spare time." Armed with a set of general instructions which he published, and with personal letters from him, they collected everything under the sun: rocks, plants, bird skins, eggs, fossils, insects, snakes, amphibians, mammals, archeological relics, tide data, anthropological artifacts.

Supplying his people with packing materials, preservatives, weapons and ammunition, he appears to have been a benevolent friend to many of them. If a collector found himself in Washington, Baird would put him up in Castle quarters, put him to work on the collections and entertain him in his home. Nor was he reluctant to pull strings in one's behalf when a co-operative amateur went out into the field. His father-in-law was Inspector-General of the Army and through him Baird frequently arranged for military men to get special assignments that benefitted the museum. If he used his influence baldly, it was for the Smithsonian's sake. When the United States Commission of Fish and Fisheries was established to investigate declining catches, he was named its first commissioner and managed to see that the Institution received the specimens it collected. When he attended the Centennial Exposition as an official observer, he returned to Washington with 42 boxcars of new material.

Regularly working 15-hour days, one year he wrote more than 3,000 letters without a stenographer. Predictably he became a nexus of natural science. As the Institution grew so did its reputation and influence, which in turn led to greater growth and more work for Baird. Private collectors rushed to donate specimens and trade duplicates. Outstanding scientists volunteered their services. There seem to have been more full-time "honorary" curators than salaried employees. Many of them held other posts simultaneously, like John Wesley Powell who both chaired the Smithsonian's Bureau of American Ethnology and directed the U.S. Geological Survey when making his great western explorations.

By the time of Baird's death, the new building was bursting at the seams with scientists and specimens. His successor, astronomer Samuel P. Langley, persuaded Congress to fund a new building in the course of his tenure which spanned two decades. Langley's successor in turn, paleontologist Charles D. Walcott who had also succeeded Powell as director of the Geological Survey, opened the National Museum of Natural History in 1911. An intentionally monumental building costing $3.5 million, it boasted a 125-foot-tall rotunda 80 feet in diameter, a Corinthian portico facing the Mall, and nearly 11 acres of floor space on four floors. (That area was doubled with the addition of two seven-story wings in the 1960s.)

In the interim an army of scientific characters had come and gone. There was a pair of entomologists who habitually stole each other's specimens and never conversed. There was a lady who grew elderly at her work and (like some others) literally lived in her laboratory whence she would emerge after hours wearing a nightgown and carrying a chamber pot. One paleontologist, so frugal that he cut up cereal boxes to use as file cards, gave up his boardinghouse lodgings each time he went on a field trip and stashed all his belongings in his office to save rent. On his death he left the Museum a small fortune. Tradition has it that Aleš Hrdlička, the Bohemian-born patriarch of physical anthropology here, measured the head of everyone who entered his office. He also built up the largest human skull collection in the world, and robbed occasional graves in the process. Another entomologist turned out to be a bigamist keeping two families in separate Washington homes. One ethnologist, long used to tramping around the West, dead-reckoned his way across town in search of a party being held in his honor, peered into several Capitol Hill houses to find the place and was arrested as a Peeping Tom. A maiden zoologist of independent means gave up her pittance of a salary if it would be used to hire a young assistant with a bride to support. The lady, Mary Jane Rathbun, worked gratis for the next 30 years and published the definitive, four-volume *Crabs of America*. The assistant, now legendary crustaceans taxonomist Waldo Schmitt, remained on the roster as a "temporary" employee for 67 years and, six years after retirement, went on yet another Antarctic trip to return with 30,000 specimens of shrimp.

Anecdotes about earlier generations of Museum scientists portray the Institution as a hotbed of people fit to be institutionalized. Some were pathologically parsimonious, others hopelessly reclusive. (One pipesmoker

Mules had not given way to Mack trucks in July, 1904, when construction for the Museum began on the Mall. The work took seven years and cost $3.5 million.

Opposite: *One of Washington's many open-air markets continued to boast a busy trade on the north side of the Museum after the building opened.*

would hide in a closet at the sound of approaching visitors and stay there while smoke seeped out around the door jamb.) Several kept their subordinates on starvation wages and demanded they work on weekends. Many refused to retire, but worked seven days a week into their 80s. Secretary Emeritus Alexander Wetmore, an ornithologist who named nearly 200 species of birds, went to work regularly after turning 90. His predecessor in the Castle, astronomer Charles G. Abbot, lived past the century mark and remained active nearly to the end.

Through all of this, a more important common denominator than eccentricity comes clear. Museum scientists have exhibited a rare degree of commitment to their work. If some of them were "maladjusted" to nominally normal life, that was because they didn't live much of it. The different drummer who kept the cadences of their careers beat a relentless meter. Their purpose, commonly, was to catalogue every known plant or animal in a chosen group; since new explorations kept coming up with new species, the work had no end. Furthermore, as it progressed according to the dogged dictates of scientific procedure, it could not be hurried, and there were no shortcuts. Each scientist's contribution was to enlarge the science he practiced. Many must have known the work would never be finished, but the magnitude of their dedication, at least, was equal to the task.

Without belaboring the Byzantine administrative history in which some scientific satraps carved disciplinary and bureaucratic fiefdoms, suffice it that the Museum has settled down as a reasonably conventional organization. Gone are the days of slave wages (though hundreds of volunteer docents, technicians and investigators still donate their services). Civil Service rules now apply and departmental chairmanships, like the Director's job, are held by tenured curators who continue their research work and return to it full-time after five years. Major departments command standard disciplinary realms. Within these reside smaller divisions. In the Vertebrate Zoology Department, for example, birds, fishes, mammals and "herptiles" (reptiles and amphibians) all have division status while in Mineral Sciences the divisions cover meteorites, mineralogy and petrology. But that's only part of the picture. Each department, and many

FARM DAIR
Chelsea

divisions, possesses fine, specialized research libraries containing rare books acquired in Joseph Henry's day. Some have seminar rooms, study carrels and laboratories for visiting scientists to augment the curators' and technicians' laboratories.

In addition to the Museum's own personnel, this building also houses special branches of other government departments which depend on the collections here: special bird, reptile and mammal sections of Interior's Fish and Wildlife Service; the Geological Survey's Paleontology and Stratigraphy Branch; the Commerce Department's Systematics Lab which researches fishes and crustaceans for the National Marine Fisheries Service; Agriculture's Systematic Entomology Laboratory which employs more insect experts than the Museum does. Many learned societies also operate out of Museum space: American Fern Society, Association for Tropical Biology, Atoll Research Bulletin, Society for Systematic Zoology, Biological Society of Washington, Entomological Society of Washington. Beyond that there are special projects and program offices for primate biology, bat banding, endangered flora, marine mammal salvage, medical entomology, Pacific vegetation, a registry of tumors in lower animals and the Scientific Event Alert Network (*SEAN*) which monitors natural ephemera like the appearance and demise of volcanic islands and whale strandings.

If all that appears a muddle, take it on faith that a certain methodology hides within this institutional madness. Mr. Smithson set no limits on his legacy's ken. From Joseph Henry to S. Dillon Ripley, the eight variously energetic, gifted and ambitious Secretaries have believed the world at large is properly the Smithsonian's oyster and its ward. To a man they have tried to make good their benefactor's goal regarding "the increase and diffusion of knowledge." Each of these scientists set priorities which reflected his own interests (and obscured the objects of his prejudice). Each Secretary left his mark by leading the Institution in chosen directions, which meant expansion of the physical plant and personnel more often than not to cope with wider concerns and more work. The original coterie became a community, then a huge corps. The Smithsonian grew to embrace 30 buildings, branches and outposts: picture palaces like the

Above: *The new building, boasting the latest in central heating and electric lights, also featured state-of-the-art exhibit cases in its regimented public halls, which admirably served the show-all-and-tell-little school of museum display.*

Right: *Behind the scenes, the new building featured the latest scientific equipment such as these rock slicers and polishers, pride of the genteel Department of Geology.*

Hirshhorn Museum and Sculpture Garden, an environmental studies center on the rustic shore of Chesapeake Bay and Washington's John F. Kennedy Center for the Performing Arts on the shored banks of the Potomac. There's a radiation laboratory in nearby Rockville, Maryland, a marine study center in Florida, a little lab on a Belize key, a tropical research center in Panama, astrophysical observatories in Cambridge, Massachusetts, and Amado, Arizona.

Like the parent organization, the Museum has grown in size and staff. Its ranges of interest have become increasingly diffused, and therein lies the Smithsonian's secret of success. Its original mandate implied a scientific center and it has kept faith with Smithson's vague vision. This descendent of the first National Museum of edifying oddities remains potentially interested in everything on earth and in the space around us, particularly the established biological sciences and all the establishments of men. Tacitly admitting it cannot do equal justice to all facets of all these things, it focuses its active curiosity through the penetrating minds of its investigators, some of them legendary deans of their specialties, others promising pioneers. The system cannot be perfect; it cannot examine every detail of earth. But by enlisting a band of scientists with a wide spectrum of interests, then giving them time and facilities to work independently, the Museum effectively expands knowledge in specific directions. The new understanding gained in specific areas is more important a legacy than the voids which inevitably remain.

Thus at the National Museum of Natural History, scientists of several sorts pursue some esoteric studies to both sublime and ridiculous lengths. Some know a single subject encyclopedically; others, like Paul Desautels who made his name as a gem scientist then switched to orchids, dare to leap from mineralogical pillar to botanical post. Some, like Dr. Frederick M. Bayer who may know more about the gorgonacean corals than anyone else, delve into one small biological group all their lives. Others, like marine ecologist and biogeographer Dr. Walter H. Adey, try to harness many disciplines with their sinuous ideas. The first curator I met here more than a decade ago prided himself on entering retirement "knowing everything about something [scorpions if memory serves] and something about everything." The best Museum scientists fulfill that hoary definition of an educated man. The worst (using the word advisedly) know everything about one thing and hardly even talk about anything else. Some spend their lives reshuffling the phylogenetic charts of every known moth or moss for reasons we'll examine later; others race about discovering obscure new animals, vegetables, minerals or concepts which become identified with their names.

It has often been said that any single scientific study resembles a brick placed in the growing edifice of human understanding. But bricks imply permanent mortar, and true scientific method assiduously tries to dismantle imperfect theories and rearrange misplaced data into stronger structures which must remain temporary. (Several Museum biologists spontaneously volunteer that their goal is "to prove Darwin *wrong*." They have in some details, such as the mechanics of "natural selection." But Darwin's outstanding idea of evolution per se withstands each new scientific test; nay it gains strength with testing.) It seems to me that a closer analogy than masonry lies in a child's connect-the-dots puzzle. As lines of understanding newly connect previously unrelated points, the outline of an image mysteriously comes into view. The picture that Museum scientists are trying to refine— each of them adding a line or merely locating a new spot—depicts the increasingly detailed portrait of our world. The picture will never be complete; it remains a vast, unfinished mosaic albeit with more pieces than last year even with the removal of a few.

Despite the systematic nature of science, the diversity of the Museum —its investigations, curators, collections and activities—seems to verge on chaos. Physically and intellectually it is a labyrinth, a place of endless passages, dead ends, cul-de-sacs. In sum, the National Museum of Natural History comprises many museums, public and restricted, visual and theoretical. There is what I'll call a "Marine Museum" in one exhibit hall and several floors of laboratories, a "Museum of Conservation" in its curatorial and international activities, a "Taxonomy Museum" in most of its biological work. The best way to explore this place of many parts is one facet at a time, because in its totality the single Museum is just about as incomprehensible as all the world itself.

04003
END SECTION
ONLY ACOMA

THE MUSEUM OF COLLECTIONS

The heart of a museum lies in its collections. Some say this Museum *is* its collections. The aggregations of these groups of objects under one roof—a literal example of *E Pluribus Unum*—seems almost as unlikely as the Institution itself. Singly and in lots weighing several tons, they have come from every imaginable kind of source via a similar variety of means including gift, bequest, permanent loan, outright purchase, even tantamount theft (in early days) and the hard collecting work of the curators themselves. In turn they've been incorporated in several ways: as singular treasures, distinctly endowed memorials, or simply as numbered elements among the 63,882,526 items in the National Collections according to one count of dubious precision. Consider:

• Jeweler Harry N. Winston gave the legendary Hope Diamond, which stands apart as the single most sought-out object in any American museum.

• Mrs. Thomas L. Casey donated 126,000 insects owned by her late husband along with a fund to preserve the Casey Collection in a room with the slight air of a shrine to it. (There's the Colonel's portrait, his captain's chair perched in perpetuity atop a bookcase, his microscope despite the tale that he wished it buried with him.)

• Meteorites Curator Roy S. Clarke, Jr., went to Mexico when the Allende meteorite fell and spread the word throughout the 200-mile neighborhood that he'd pay cash for fragments of the celestial stone. He returned with two tons of them.

• Leonard Stejneger, whose half-century career here began in 1889, acquired a number of important turtles by swapping away other specimens as if they were his own and left a legacy of raised eyebrows.

• Some great nineteenth-century anthropologists are reputed to have stooped to stealth, seduction, force and simple thievery to bring home the ethnological bacon.

• In the Spencer Baird tradition, Ronald W. Hodges gave the Museum 25,000 moths which he'd collected himself when he came here to join the Systematic Entomology Lab.

• Beetleman Paul J. Spangler, through trades, purchases, persuasion and his own field work—which accounts for nearly 600,000 specimens—has added 1,100,000 insects to the lot.

Hundreds of Zuni pots line the Museum's attic shelves. This late-nineteenth-century collection, considered priceless by anthropologists for the information it reveals about Native American culture, has played a role in a recent renaissance of Zuni ceramic art.

Opposite: *The Hope Diamond, most celebrated object in the collections, has a history less bizarre than the myth of a curse to avenge its theft from a heathen idol's eye. In fact, a reputable merchant brought a huge, flawless blue diamond out of India about 1642 and sold it to Louis XIV. One of the crown jewels of France for a century and a half, this was stolen during the Revolution and apparently recut. Thirty years later a London dealer sold the largest surviving fragment to Henry Philip Hope, an English collector whose heirs owned it for two generations. After changing hands, it was purchased by Paris jeweler Pierre Cartier, who spun fabulous yarns to intrigue clients. American heiress Evalyn Walsh McLean bought it on the installment plan in 1911 and wore it with gay abandon for four decades. She sold it to Harry Winston, who gave it to the Smithsonian in 1958. Weighing 45.52 carats and measuring 25.6 by 21.9 by 12.05 mm, this gem is remarkable for its deep color, rare size and flawless clarity.*

• Aleš Hrdlička, a legend still haunting these halls, assembled one of the largest bags of human bones in Christendom. Though it contained some 20,000 human skulls, he thought there was always room for more and urged colleagues to leave their heads to the Museum. (His own postmortem instructions dictated cremation. His ashes were mingled in an urn with those of his first wife and placed in his old lab—to his surviving widow's dismay.)

The frenetic collecting tradition, itself something of a legacy so far as the Museum is concerned, began with the U.S. Exploring Expedition. That odyssean field trip resulted in the seminal and pre-Smithsonian Wilkes Collection, a priceless pot pourri and something of an administrative fiction in that its elements have been scattered throughout the building. One of the oldest surviving group of objects, it was aggregated late and dispersed early to form the nuclei of many collections here. The Wilkes material might have been anticipated before Smithson's death, was only acquired by political happenstance a decade after the Institution's founding, and like the Smithsonian itself owed its genesis to John Quincy Adams.

As President, Adams had encouraged Congress to launch a voyage of exploration and discovery across the Pacific for several purposes: to chart the waters and their navigational hazards; to discover islands, fishing grounds, trade routes; to push back the boundaries of *terra (et aqua) incognita*. Diverse supporters lobbied for the idea. (Yankee merchants and whalers, for example, anxious to learn more about the western ocean at anyone else's expense, joined forces with a singular eccentric who meant to shorten sea lanes by sailing inside the earth through the preposterous features of his "Holes in the Poles Theory.") In 1829 Congress approved the expedition in principle, then began nine years' wrangling over details before Lt. Charles Wilkes embarked from Hampton Roads in the van of a six-ship Navy squadron. By coincidence he set sail just four weeks and a day after Smithson's gold guineas were minted into American Eagles.

Wilkes' goals, hammered out by Congress and refined in the Navy Secretary's orders, were to advance "the great interests of commerce and navigation," and when compatible with that to "extend the bounds of

Opposite: *The Museum's collections of stones include many nameless gems of unusual beauty. Weighing between 39 and 1174 carats, this arrangement includes (clockwise from upper right): greenish-blue aquamarine, citrine, heliodor, blue aquamarine, clear topaz, reddish topaz, malaya (a variety of garnet) and round blue topaz. The malaya is from East Africa, the clear topaz from West Africa and the rest from a single Brazilian state, Minas Gerais.*

Left: *The Indian Emerald necklace comprises two dozen tumbled green gems—the largest of them 25 carats—and four dozen beads in a modern platinum setting encrusted with diamonds. It was donated to the Museum by Mrs. Marjorie Post.*

Overleaf: *Mimetite crystals can serve as a source of lead, but the Museum treasures this specimen for its delicate beauty.*

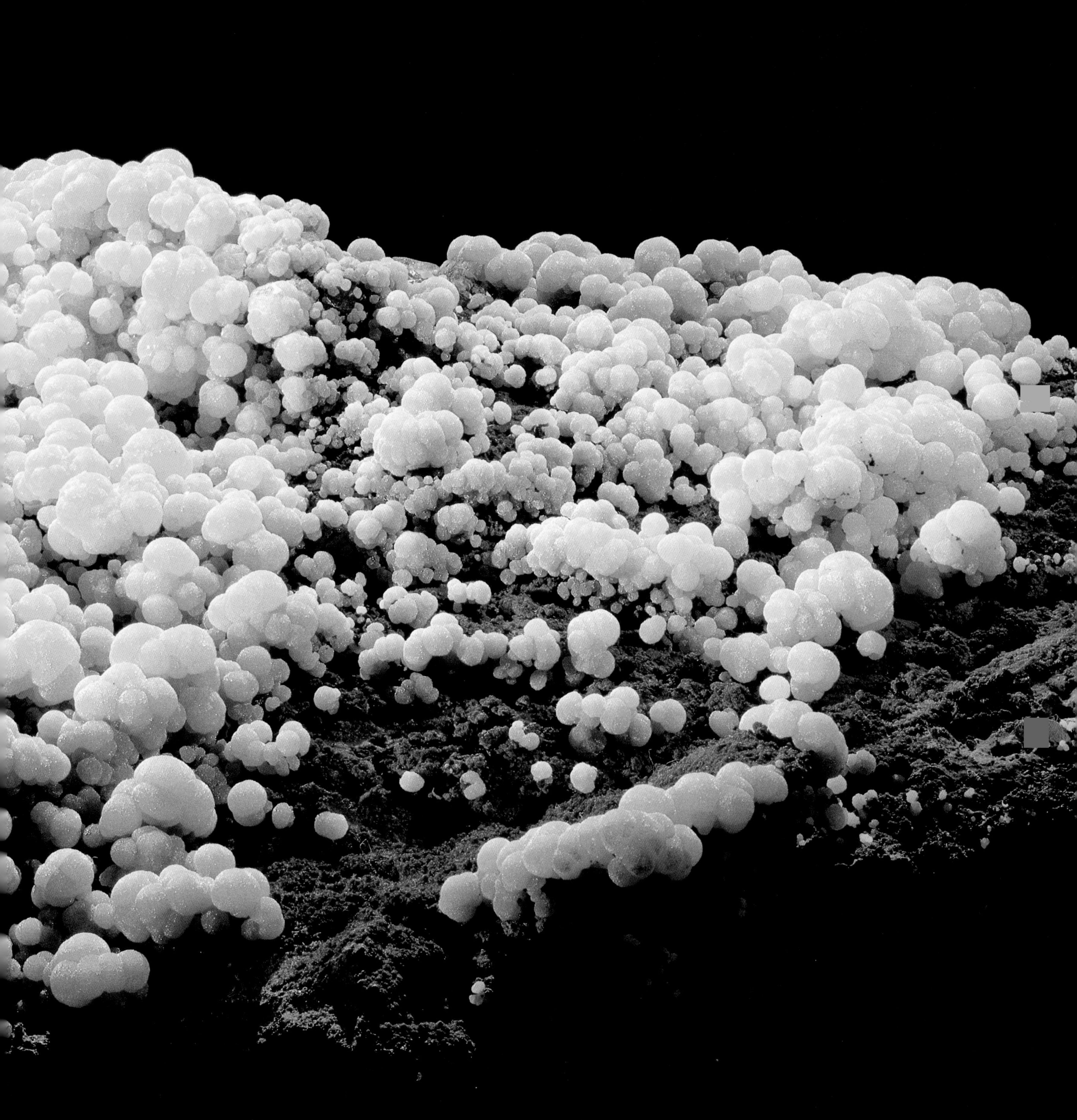

The anthropology collections include a pair of ear pendants (right) *made of iridescent beetle wing covers and feather tufts by the Jívaro of Ecuador. The same people made the dance cap of toucan feathers—white from the breast, orange and yellow from the tail* (opposite).

science and promote the acquisition of knowledge," a watchword for those times. The scientists who sailed in the specially refitted ships were not bright amateurs and therein lay a signal distinction of the U.S. Exploring Expedition, according to Herman Viola, who heads the National Anthropological Archives within the Museum. Gone were the inquisitive men of leisure who had previously dominated the sciences; this project was the first to be manned by professionals.

The geologist aboard was James Dwight Dana, author, Yale University professor and "one of the greatest American leaders of science." (His classic texts on mineralogy remain broadly authoritative to this day.) Linguist and ethnologist Horatio Hale, 21 when they embarked, nearly missed the cruise because Wilkes believed any simple literate could "pick up languages well enough to make vocabularies." Titian Peale, who filled a berth as zoologist and artist, was portrait painter Charles Willson Peale's younger son and sometime manager of the family's natural history museum in Philadelphia, the first public (and notably bizarre) attraction of its kind in the land. Scotsman William D. Breckenridge, a distinguished botanist, was almost barred by Wilkes because he wasn't American. The services of two other outstanding investigators were terminated early, in one case because the scientist refused to turn over all his notes and drawings to the

Despite its enormous size, the building lacks storage space and collections managers find some use for almost every cranny, like this dusty loft below the roof. Stuffed mammals from bygone exhibits abide here with perfect patience.

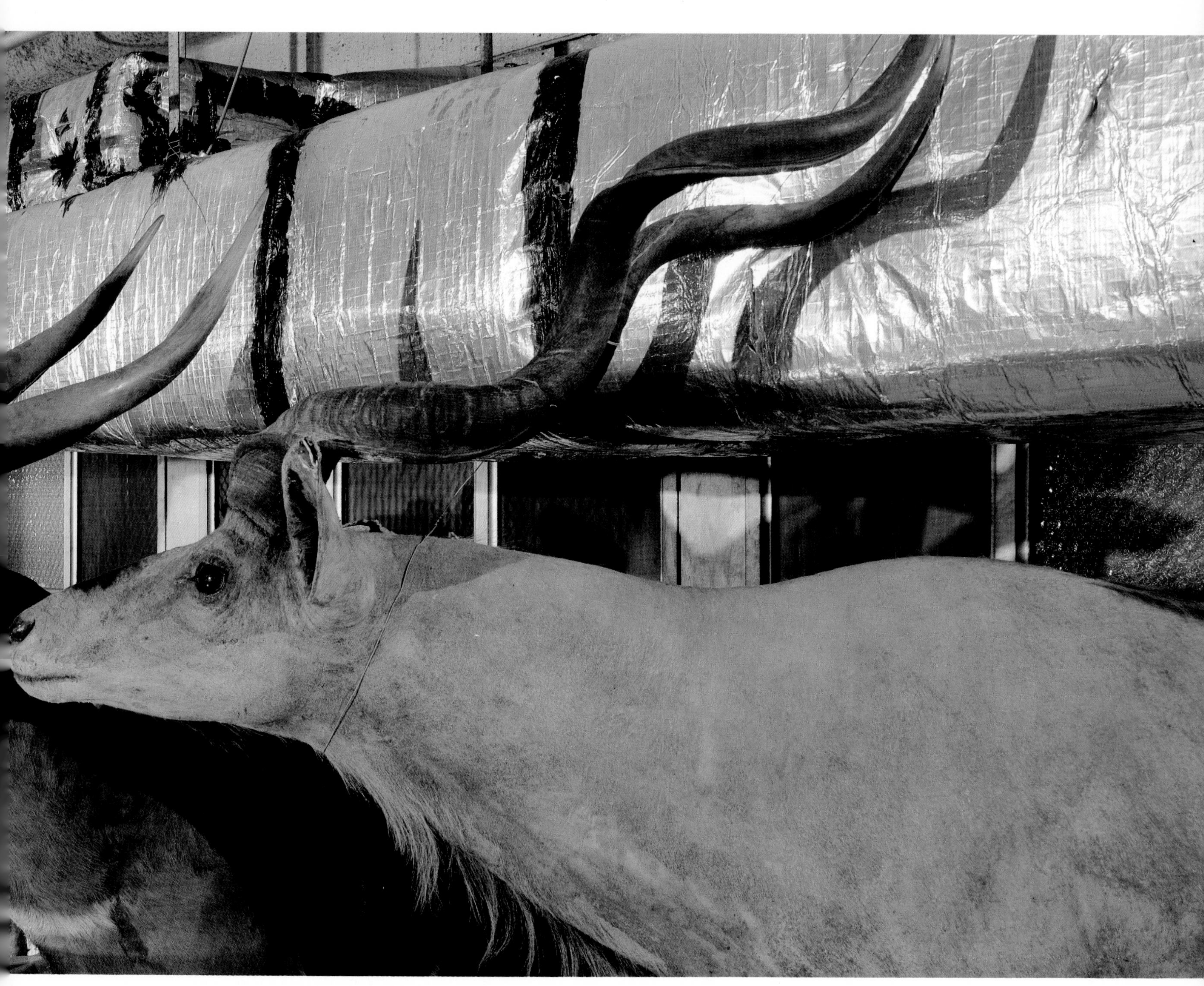

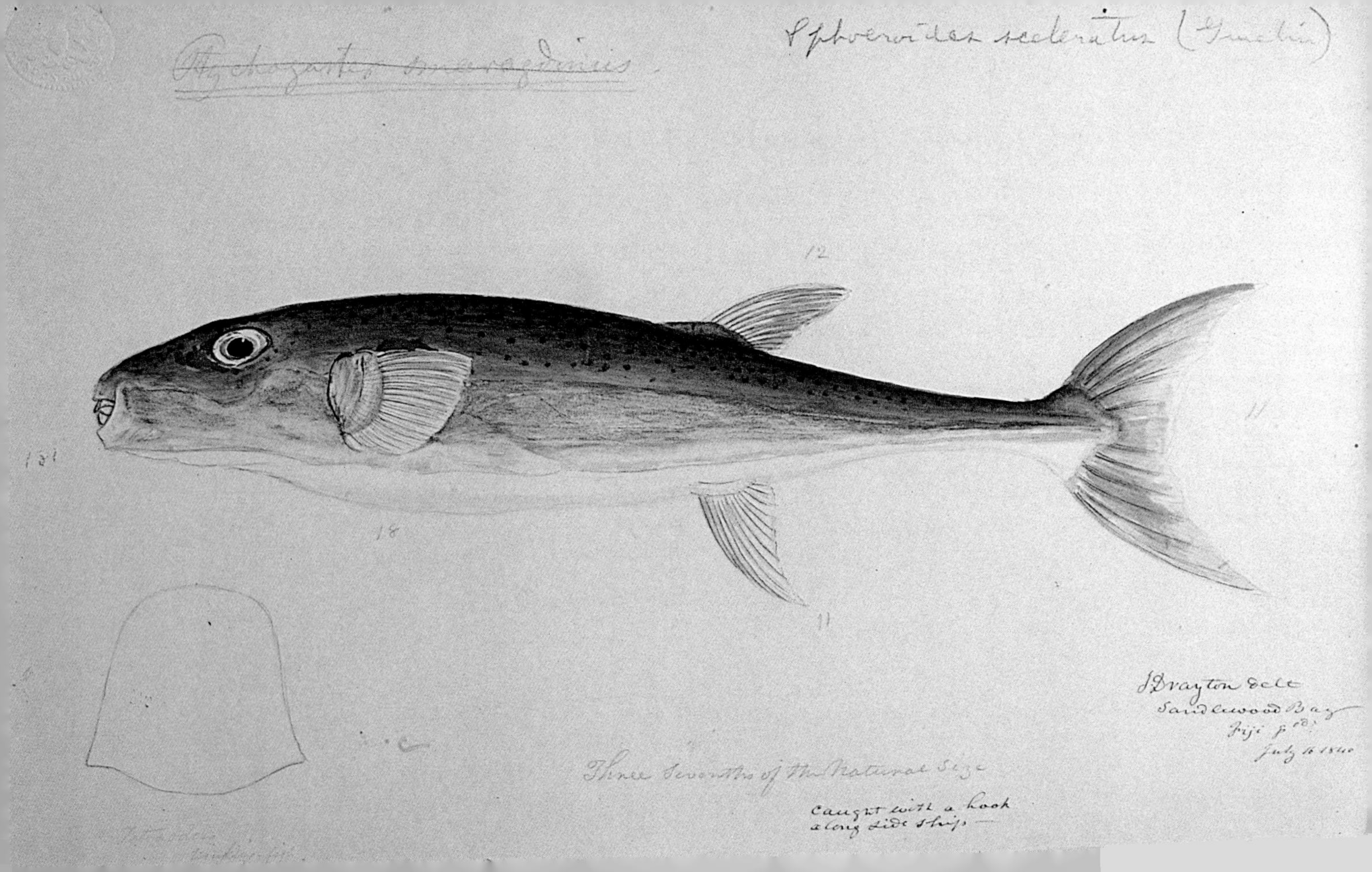

~~Ptychogaster smaragdinus~~

Sphoeroides sceleratus (Gmelin)

12

131

11

18

11

J. Drayton delt
Sandlewood Bay
Fiji I^{ds}
July 16 1840

Three Sevenths of the Natural Size

Caught with a hook
along side ship —

Left: *When* Tetradous diodous *was caught with hook and line in the Fiji Islands, artist William Drayton made this rendering, one item among thousands of drawings and specimens that became part of the nominal "Wilkes Collection." That aggregation was later turned over to the Smithsonian and broken up to become the nucleus of many collections within the National Museum of Natural History.*

The engraving of Lt. Charles Wilkes graced the frontispiece of his official five-volume narrative of the 87,000-mile, almost 4-year-long United States Exploring Expedition.

fleet commander. Officer or scientist, if a man crossed Wilkes he might be assigned to the convoy's supply ship and sent home. Erring crew members fared worse.

An egomaniac of the first water, Lieutenant Wilkes was a martinet, a biological ignoramus and a cat-o'-ninetails disciplinarian. (In this last respect he was not unique, only extreme. Though a dozen lashes was the prescribed punishment for breaches of Naval discipline, he often ordered 40 strokes of the whip. He was later courtmartialed for flogging three Marines every other day for a week as an incentive for them to reenlist.) Fully a quarter of his men deserted during the four-year cruise and his officers detested him. But for all his foibles, he got the job done. Indeed, scholars now believe that "the secretive and aloof Wilkes" provided Herman Melville with a model for Captain Ahab in *Moby Dick*. (That novel profited from this expedition in other particulars as well. The drawing of a tattooed Pacific islander became the prototype of Queequeeg.) Dana offered this balanced assessment of the Commander: "Wilkes, although overbearing and conceited, exhibited through the whole cruise a wonderful degree of energy and was bold even to rashness in his explorations. . . . I much doubt if, with any other commander that could have been selected, we should have fared better."

In many respects an epochal scientific and maritime success, the voyage was also a tragedy of errors punctuated by occasionally absurd comic moments. Once when the eminent conchologist Joseph P. Couthouy was off collecting specimens, a mess boy cleaned up his cabin and threw all the sea shells overboard. When a vessel foundered off the Oregon coast, all hands were saved, but Peale's butterfly collection went down with the ship. On some islands Wilkes would only permit the scientists an hour or two ashore—apparently in the spiteful belief that his own navigational feats would eclipse their discoveries if he could keep their work within some bounds. But he undermined them in other ways as well. When scientists planned to pay crewmen for unusual specimens they found ashore, Wilkes announced that everything collected was government property and the crew gave up even trying to help.

Some blots on the Expedition's record resulted from human error;

The flax cloak now gracing the mannikin in the foreground of this Maori diorama was collected by Wilkes in New Zealand. Illustrations and information gleaned from official accounts of the Expedition may have provided Herman Melville with models for characters in Moby Dick *such as the tattooed harpooner Queequeeg.*

others reflected the huge perils of the time. When two officers thought they spotted land in the high latitudes on January 16, 1840, they failed to log the sighting. Three days later they made another landfall, according to dubious logs replete with erasures, just hours before a French ship also claimed discovery of Antarctica. When bad weather broke up the convoy off Tierra del Fuego, one ship was lost without a trace. Finally, hopes to explore the interior of many Pacific islands were dashed by the unmistakable hostility of the natives, some of them cannibals. (Wilkes' own nephew and another officer were killed in one battle with islanders.) In the Fijis, Wilkes went ashore armed with the warrant for the arrest of a chief named Vendovi who had murdered ten American sailors several years earlier. The suspect turned out to be the man they'd hired as a pilot to steer them through uncharted reefs. He agreed to come along peaceably enough, evidently looking forward to seeing America. But after two years as a prisoner aboard ship, he fell ill and died hours after the voyage's end.

Wilkes' ship arrived home first—after he'd dispatched the other vessels in the squadron on a fool's errand. But the lieutenant was disappointed at not receiving a hero's welcome. Instead there were hearings, courts-martial, and some slapping of wrists (notably his own) before he took personal charge of the specimens and sat down to write a five-volume official history of the Expedition. It recounted in humorless and mind-numbing detail the voyage's substantial accomplishments. Wilkes had crossed the Pacific three times, circumnavigated the globe, sailed 87,000 miles, drawn 200 useful and remarkably accurate charts, and established the continental proportions of Antarctica, part of which bears the name Wilkes Land today.

The scientists' collections of specimens were placed in the care of Poinsett's struggling National Institute and displayed in the Great Hall of the old Patent Office Building (now the National Portrait Gallery and National Museum of American Art). Titian Peale, who was later hired as curator, blanched at what happened there. Congressmen, amateur collectors and passersby helped themselves to curiosities. Bird skeletons were hopelessly mixed in boxes. Hundreds of insects, gathered throughout the Pacific expanses, were jumbled together so that information of their

COLLECTIONS
of the
UNITED STATES SOUTH SEA SURVEYING
and
EXPLORING EXPEDITION
1838, 9, 40, 41, & 42.

Ethnological specimens 2487 yellow numbers
— " — duplic. nos. 29 total number of specimens. 2516
Mammalogical species [illegible] red numbers.
total specimens, — — 134
Ornithological species 762 blue numbers
total specimens. — 2150
~~4800~~

By T. R. Peale
U. S. Patent Office
1846

The collections made by the Exploring Expedition under Capt. Wilkes, were brought to Washington and unpacked at the U.S. Patent Office, where they were on exhibition for several years. About 1857 they were transferred to the Smithsonian Building. This Catalogue was presented to the Natl. Museum — Jany 1877 by Mr Peale in person.

of the Brig Charles Daggett, Captn Bachelor, of Salem in 1833. He confessed having plotted the destruction of the crew, and the plunder of the vessel; and succeeded in killing ~~but~~ ten of the men, who were cooked and eaten; the rest escaped with the vessel. He died at New York in June 1842.

31 to 35. Crania of Fiji Islanders.

36. Head of Tekea, of Taranaka. A chief of the Notekesookee tribe, New Zealand.

37. Head of Atee, of Wanganewe; Notepahea tribe, New Zealand; these heads were dried in ovens, and according to the custom of the country, were kept as trophies of victory, by the enemies of their tribes.

38. Cranium of a Chinook chief, from Nesqually, Oregon; it was flattened artificially during infancy, according to a custom of that tribe; a model made by the natives of the cradle in which this operation is performed may be seen by referring to number 116.

39 to 55 Crania from dead Island, and other places on the lower part of the Columbia river, Oregon. All but numbers 47 & 52, are more or less flattened by artificial process

Titian Peale, son of the famous portraitist, catalogued the anthropological material collected by the Wilkes Expedition. The catalogue is now in the National Anthropological Archives.

Wilkes himself drew a seascape in Disappointment Bay, Antarctica, while seamen and scientists variously relaxed, filled ship's kegs with fresh water and gathered rocks and small organisms from this ice island. When Herman Melville saw this published engraving of the drawing, his imagination seized upon the appearance of the weathered floe. It looked to him like a white whale and reportedly inspired the central image for Moby Dick, *the novel which borrowed many details from accounts of Wilkes' scientific odyssey.*

Another engraving from an official report depicts Pacific Islands women making tapa cloth. Wilkes returned with many examples of the material, some of them unique, which grace the exhibits today.

Opposite: *James D. Dana collected hundreds of previously unknown corals during the Expedition, including* Corallium secundum, *one of the so-called "precious corals" which jewelers prize. Dana's specimens, and the illustrated monograph describing them, remain in the Museum's peerless collections of marine invertebrates.*

origins was lost. The scientific value of many of Couthouy's mollusks was similarly destroyed; metal tags identifying each shell discolored the alcohol preservative, so someone took all the tags and set them in a jar by themselves. Most outlandish, perhaps, a collection of animal fossils was consigned to a Georgetown bone mill, ground up and sold for fertilizer.

Clearly the remaining material could not stay where it was and Congress decided it should reside in the Smithsonian, over Joseph Henry's vain objections. Despite his misgivings, the Smithsonian received a wealth of material, for not all of the Expedition's hard-won specimens had been lost. Not by a long shot. As a result, today the Museum possesses countless specimens of inestimable value because of their age and provenance. Dana collected 483 species of corals; 229 of them had never been described before. His unique specimens became the heart of the world's largest coral collection, one that represents a virginal sample of marine fauna impossible to collect today. Breckenridge, who became superintendent of the Capitol, brought back 600 new plants which became the core of the Botany collections; the seeds he gathered formed the nucleus of Washington's Botanic Gardens. Of the more than 500 species of fishes that the voyagers collected, nearly 200 were then unnamed and unknown to science.

In addition to advancing baseline biology, Wilkes' odyssey gathered priceless anthropological material as well. The vocabularies and grammars which Horatio Hale prepared have never been surpassed. (Some have become inestimably valuable to linguists studying the phylogeny of speech because they contain the best record of languages now dead.) Less academically, native armor collected by the explorers now acquaints millions of visitors with an antique ethos; it is displayed along with a

xploring Expedition

ZOOPHYTES,

1

3. a.

1. a.

3

2

Fiji Islands chief Vendovi, named in a warrant for the murder of American sailors, willingly became Wilkes' prisoner, apparently in the vain hope of seeing America.

Opposite: *Custodian of many mementoes received by statesmen as official gifts, the Museum finds some of them worthy of display. Several of Japan's most eminent armorers made components of this suit of Samurai armor over three centuries. The helmet bowl, for example, was made in the early sixteenth century, the face mask in the nineteenth. The Emperor of Japan presented the whole suit to President Roosevelt after the Portsmouth Peace Conference settled the Russo-Japanese War in 1905.*

warclub found beside the body of Wilkes' nephew. Among specimens which grace the botany, zoology and anthropology collections, the most singular may be Vendovi's skull. The adventurous chief wanted to reside in America; his cranium, carefully preserved, still does.

The acquisition of the Patent Office material, which in addition to the Wilkes Collection contained some of Lewis and Clark's souvenirs from 1804, seems fortuitous in retrospect. A team of scientists went abroad for good purposes and the palpable fruits of their labor happened to fall in the Institution's infant lap. Once the Smithsonian started to mature, it developed an insatiable appetite for collecting things and intentionally built its collections in various ways. Innumerable expeditions were setting forth in those decades for diverse military and civilian purposes. The Canadian and Mexican boundaries were surveyed; wagon train and railroad routes were traced. Each party garnered new material and sent it to the Smithsonian, in part because the Institution's mandate seemed to require it, in part because Henry and Baird outfitted members of these parties and armed them with instructions. The secretaries also equipped individuals to gather natural history material, a genre that included objects found in archaic Indian mounds and inspired the Institution's embrace of de facto anthropology. One singular expeditionist was John Wesley Powell, later head of the Geological Survey and Bureau of American Ethnology simultaneously. Though he'd lost an arm at the Battle of Shiloh, he still had the capacity to make the maiden white water trip down the Colorado River. His verve and breadth of interests typify the Institution's spirit. Nothing was beneath notice; few obstacles were impassable; and examples of just about everything under the sun were subject to inspection and preservation. Meanwhile, the collections multiplied like rabbits.

The question arises, what are these collections good for? One might as well ask the purpose of a library's books, a church's birth, baptism, marriage and death records, an accountant's ledgers, a legislature's proceedings, an abbey's chronicles. The Museum's biological and mineral collections comprise physical records of the natural world's changing states; the anthropology collections serve as cultural time capsules. Some cases in point:

Below: *A physical catalogue of many cultures around the world, the Anthropology Collections contain a vast array of North American Indian materials. The Pomo Indians of California, unusually talented wicker workers, made these gift baskets adorned with feathers. They used a variety of materials—willow shoots, sedge roots, redbud bark and more—in their basketry, which provided containers for every purpose. (Some were woven so loosely they caught fish, some so tightly they held water set on the boil with fire-heated stones.)*

Opposite: *The Alaskan Eskimo mask, probably worn during ceremonial dances, features a seal's flippers, skin and face along with the grinning visage of the animal's semi-human* inua, *or spirit. The object holds clues to Eskimo mythology and religious beliefs. Its goggled eyes indicate that the* inua's *identity is hidden and inscrutable. A fringe of caribou fur surrounds the mask.*

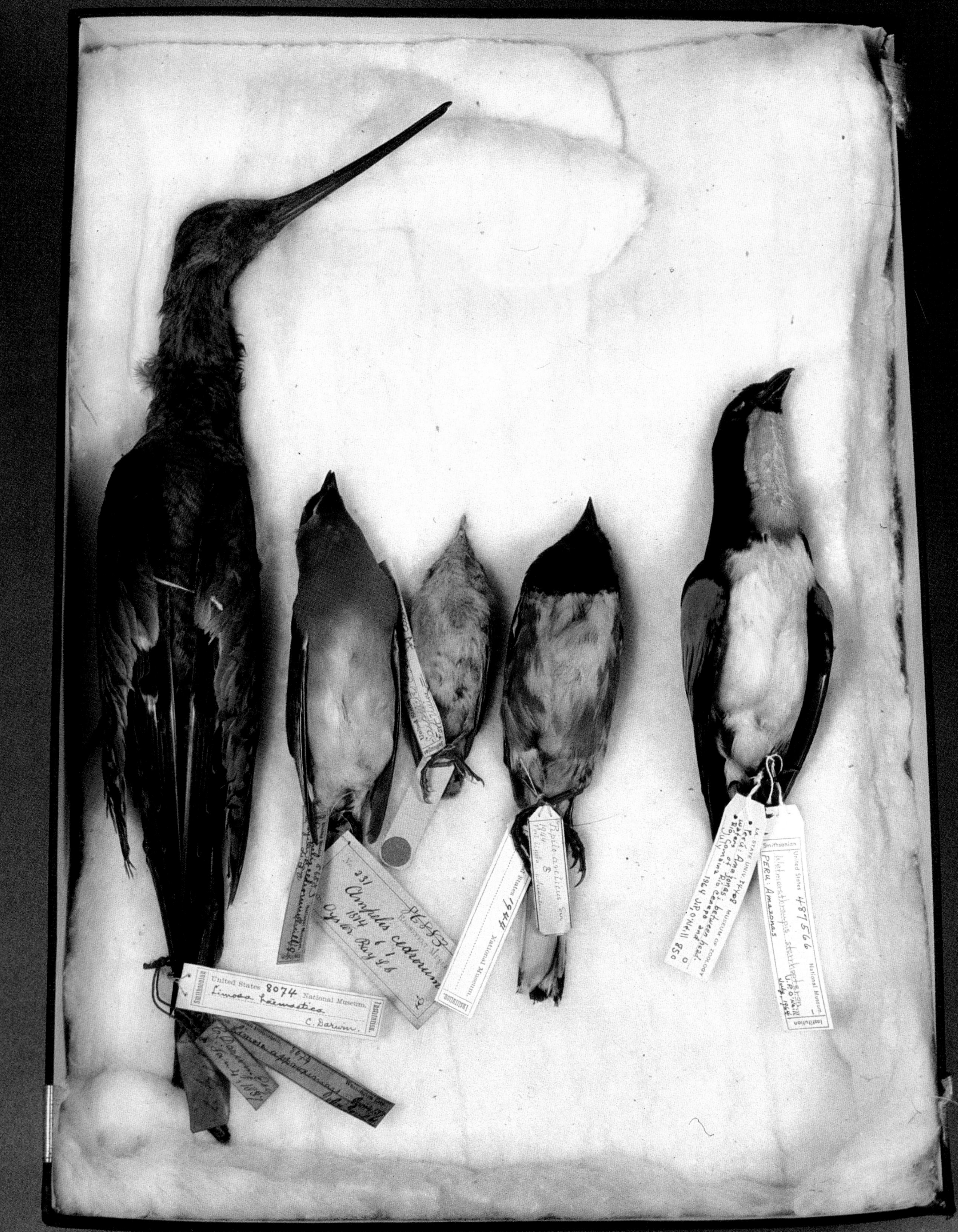
United States 8074 National Museum.
Limosa haemastica
C. Darwin.
Smithsonian Institution.
Ampelis cedrorum
1874
Oyster Bay L.I.
487566
PERU Amazonas

By our current lights, Charles Emil Bendire, an Army surgeon, seems an acquisitive eccentric. The man for whom a subspecies of the red crossbill was named, Bendire served as Honorary Curator of the Department of Birds' Eggs (a.k.a. Section of Oology) from 1884 until 1897. Collecting some 10,000 specimens with and without nests, and leaving eight cubic feet of papers, he died in unremunerated harness. Several decades later raptor populations went into precipitous decline and his work would reveal why. The discovered cause was DDT and other persistent pesticides which change to toxic DDE after ingestion by an insect that is then eaten by a flicker and on up through the food chain. In peregrine falcons, which eat other birds almost exclusively, and in fish-eating ospreys the chemical interferes with parathyroid function which governs calcium metabolism. Adult females began laying eggs containing too little of the essential element. Commonly the shell, an embryo's major source of calcium, simply failed to supply enough for healthy bone growth. In many recorded cases, the shells were so thin that they couldn't survive the rigors of incubation. In either case, otherwise healthy wild birds failed to hatch and rear healthy offspring; they kept on mating but so far as posterity was concerned they might as well have been sterile, celibate or both. This calcium scenario made physiological sense, but remained to be proven until Fish and Wildlife Service ornithologists examined Bendire's eggs. The Museum's collection contained a statistically valid sample of nineteenth-century shells which measured fractions of a millimeter thicker than eggs of the DDT generation—enough to make an organic difference to their embryonic occupants.

Similarly and more happily, the mercury-in-swordfish scare was defused though Museum collections. Fisheries authorities stumbled upon what appeared to be dangerously high levels of heavy metals in table-bound catches. They leapt to the conclusion that industrial pollution was contaminating these fish and banned them for human consumption. The move proved a hasty one. Specimens in the Fishes Division collections were assayed and found to contain approximately the same amounts of the toxic substances. Folks ate swordfish aplenty in the nineteenth century without statistically significant aftereffects. The ban was lifted.

Preceding pages: *The Zuni Indians of western New Mexico became justly famous for their decorated household pots. Zuni women built up coils of wet clay strengthened with crushed pottery fragments, then shaped and smoothed the vessel. After drying for a day or two, the pot was covered with a thin slip of kaolin clay, then polished. Using a yucca leaf brush, the women adorned each pot with a design in black and yellow ocher, which turned red when the piece was fired. This example, emblazoned with a deer, shows the animal's vital "life line," its throat, stomach and heart—thereby revealing the deer's spirit.*

Opposite: *As the labels show, the bird collections contain these historical curiosities. Left to right: a Hudsonian godwit collected by Charles Darwin in 1837, a cedar waxwing which Theodore Roosevelt collected as a teenager, a rock wren and rufous-sided towhee which John James Audubon gave to Spencer Baird, second Secretary of the Smithsonian, and an orange-throated tanager of a genus named for sixth Secretary Alexander Wetmore.*

Following pages: *Two dozen empty shells from what was called the Oology Collection include the smallest found today, a hummingbird egg the size of a vitamin pill, and the largest, an ostrich egg big as a football. Mottled patterns, acquired during the laying process, often serve as camouflage when parents leave the clutch. Sharply pointed eggs—which cannot roll far—belong to seabirds that nest on narrow ledges.*

Right: *Bird nests in a collection contain information of unpredictable utility for science. Elsewhere, the labels attached to nestless specimens reveal something of the Institution's human connections. Birds collected by Charles Darwin, John James Audubon and Teddy Roosevelt all abide here along with a member of a genus named for Secretary Wetmore.*

Right, below: *The collection features this oddity: an eastern phoebe's nest containing the skeleton of its original builder and occupant.*

Opposite: *The jar's label indicates that these crabs were taken by the Fish and Wildlife Service research ship* Pelican, *which surveyed southern waters in search of shrimp to expand the fishery before World War II.* Pelican's *log provides essential data: Station 170–2 lay at 28° 33′N, 80° 05′ W or 28 miles east northeast of Cape Canaveral in 35 fathoms. A future trawl may produce larger or smaller crabs of the same species, more or fewer of them, perhaps entirely different animals—for reasons that then invite discovery. Tissue comparisons may show a change in chemical environment while later studies of old material can expand biohistorical knowledge. The kinds of information that will be deduced depend largely on what questions scientists ask; the Museum preserves such specimens so that new questions can be asked—and answered.*

Call it the Serendipity Coefficient, an uncertainty factor underlies the potential future value of all the Museum's collections, even ones that seem accidental aggregations. No one can anticipate just how their obscure and dusty contents may one day prove critically valuable to posterity; these relics contain raw information—answers to questions that haven't been asked yet. As Assistant Secretary for Science David Challinor says potentially of every item in the collections, "You'll never know when you need it." As Secretary Ripley has written "Today the state of the planet is a major source of concern. We are entering a gigantic 'countdown' with no easy way to measure the time. Monumental changes in Earth's tropic zones, and subtle changes in its marine environments, are occurring. They may produce irreversible effects on the ability of the planet to support what we consider to be a normal pattern of sustainable human life. . . . Natural history collections serve inevitably as data banks, figures marked upon

BAIL HERE
Pelican
Station 170
Try net

Opposite: *Braced for long-term storage and future study, whale skulls dwarf a Vertebrate Zoology Department collections man.*

Following pages: *Cuban tree snails like these the size of escargots—all members of a species treasured for its varicolored beauty—face an uncertain future. Neither pests nor delicacies, they are threatened by commercial collectors, who gather too many for use as ornaments.*

the clock face of environmental time." From them we can learn the states of things past, judge the present more wisely and more clearly anticipate the future—perhaps even in time to change it for the better.

For every biologist, the most fundamentally important collections are the aggregations of "type specimens." A holotype or "type," as it's known in the trade, is the single preserved specimen which a taxonomist described in meticulous detail and entered in the biological literature with a properly-coined scientific name. Preserved and tended with special care, this name-bearer for the species remains available as the standard by which all others may be compared. (Among zoologists paratypes are specimens of the same species taken from the same original batch. An allotype is a specimen of the opposite sex from the holotype.) The Museum shelters many thousands of these specimens and they include its most valuable biological material. During an air raid drill in early World War II, several senior scientists ran for the exits with armloads of jars distinguished with red labels or ribbons. Later many of the types were transferred to underground tunnels, receiving the kind of special treatment otherwise reserved for objects like the Declaration of Independence, a type specimen of a different order.

The Springer Collection, Porter M. Kier's special charge in the Paleobiology Department, was bequeathed by a railroad lawyer who turned the boyhood hobby of digging up fossils around his Iowa home into an adult passion. A gifted amateur in the great Victorian tradition (and author of some reputable monographs), he gathered a rare variety of organisms from the Paleozoic and Mesozoic eras, especially crinoids and other echinoderms, the group that includes modern spiny sea urchins and sand dollars. Augmenting these fossils with missing links of established ages found in a few other museums around the world, Dr. Kier has been able to reconstruct the surprisingly swift evolution of a few species in a study that illustrates natural motives for evolution. Time was when an ancestral echinoderm lived well enough atop the marine mud and selected bits of organic food from the water. But the increase and diffusion of predators brought new difficulties for weak things that lived in the open. Prodded by the stick of predation and attracted by the carrot of a new food supply,

The mollusk collections contain three related scallops: Japan's smoothly fluted noble pecten, the calico shelled mantle scallop from a South Pacific archipelago and the larger lion's paw of Florida waters.

Opposite: *Found in South Dakota shale deposit and preserved in the Museum, this 80-million-year-old fossil ammonite displays a mother-of-pearl shell. Paleontologists studying these once numerous and diverse animals get clues to their behavior and biology from specimens of their single living relative, the chambered nautilus, found in the Invertebrate Zoology collections.*

Overleaf: *A slab containing two kinds of 160-million-year-old cephalopods was selected from a fossil collection for their unusual beauty and specially prepared for the popular "Splendors of Nature" exhibit.* Perisphinctes, *members of an ammonite genus, curl like rams' horns. As do their chambered nautilus survivors, in life they swam base downward oriented like the large example in the lower center. The polished objects, shaped like plastic pens, are the internal stiffeners of* Belemnites, *kin of the modern squids.*

some of these animals entered the substrate to hide in a habitat containing a mix of inorganic dirt and nutritious particles. However, it took considerable changes in anatomy to make the move permanently. The anus, for example, started out on top of these pincushions, an appropriate place when the animal lived in moving water which wafted away its scant excreta. But when it colonized the mud, the animal began unselectively ingesting huge amounts of material and passing it through its changing gut. In the process its anus moved aft and down so that now the animal left a trail behind it rather than smothering itself in wastes. Overall body shape and breathing apparatus changed too during the move. "In only ten million years," Kier writes, one group came to "possess all the features necessary to . . . live buried in the sediment."

It took the Springer Collection to provide the backbone of proof that these organisms evolved with surprising speed. Comparing its specimens with those collected elsewhere, Kier could determine that there were fewer intermediate steps than other investigators had supposed. "This rapid evolution and diversification seem to result from a sudden adaptive breakthrough. The evolutionary steps must have been large"—not a long series of minute changes as Darwin and his unreconstructed followers assumed on the basis of fragmentary evidence. Given the nature of the fossil record, all evidence is intrinsically incomplete, but a seminal collection that has been augmented over the decades and systematically compared with other lesser collections can provide as full a picture as the paleontological world contains.

The Burgess Shale Collection, another Paleozoic gold mine, contains the black-on-black fossils that then-Secretary Walcott wrested from a single quarry in British Columbia. This place was a sea some 600 million years ago until a rare combination of events first swiftly turned the water stagnant, then preserved its contents in shale. Scavenging stopped in an instant; even decay-causing bacteria perished. The result: a catalogue of the biota within one Precambrian habitat, nearly a yearbook of its denizens all illustrated by their physical remains. Recently the quarry was reopened for further excavation, but no new animals of major importance were unearthed. "Walcott found it all," Kier chortles, and his heirs are still

Specimens in the Terry Collection.

describing it. By the same token, catalogues of other times were found in Texas' Glass Mountains and Wyoming's Green River. These collections constitute other unexpurgated volumes in the history of life.

Even after a collection has been sorted and described, it remains a rich resource for further scientific study and enlightenment. Variously dusty, drab, chaotic, colorful or even macabre, systematic aggregations of objects contain raw information. Their most obvious value lies in the fact that they are, in the Secretary's cybernetic idiom, data banks whose bits can be recombined to answer new questions. The Terry Collection, for example, contains some 1,500 human skeletons along with the medical histories of their original possessors. This material was originally compiled at the Washington University Medical School. When the St. Louis institution decided to dispose of it—the cabinets stood in a planned doorway to a new wing—T. Dale Stewart (a curator and medical doctor) leapt at the chance to acquire it for the Museum on a permanent loan basis. This statistically significant sample allows physical anthropologists to correlate skeletal characteristics with known ethnic, age, environmental and disease factors. Thus peculiarities found in the skeletons of unknown people, whether ancient or recent, can be diagnosed to reveal new information about their lives or cultures. An archaeological site containing a cemetery can reveal longevity patterns, indicate what kinds of diseases and injuries the community survived and which ones decimated it, suggest how long they lived and much more. The key lies in having a group of well-documented human skeletons for study and comparison—a collection.

In the more purely cultural realm, the Nelson Collection was gathered by and named for Edward W. Nelson, an Army Signal Corps officer whom Baird persuaded to explore Alaska's Bering Sea coasts in the 1880s. He spent three years traveling from hamlet to hamlet, encountering Eskimos who until then had very little contact with Europeans. "Nelson didn't know anything about anthropology," curator William W. Fitzhugh told me while organizing the first show to use the material extensively. Unburdened by any scientific bias or the desire to prove some pet thesis, he devoted his immense energy to collecting anything and everything he could carry away with him: ivory needles and bone cases, weapons, painted cooking utensils,

Above: *The Incas commonly performed trephination—the surgical removal of skull bone—with stone instruments but neither antiseptics nor painkillers as far as we know. Hundreds of Museum specimens prove that Pre-Columbian men, women and children survived the procedure after suffering potentially fatal head wounds. (In this case, however, lack of new bone growth indicates the patient died soon after the operation.) Trephination was performed in many parts of the ancient world, including Europe and the Near East. It is still practiced in parts of modern Africa—again without drugs—to treat emotional problems and headache.*

Right: *Visiting scientists study specimens in the Terry Collection, which comprises skeletons of known individuals along with their medical histories. Thus this priceless resource allows physical anthropologists to correlate once-vital statistics with the kind of physical remains most often preserved—in ancient cemeteries, for example—and thus deduce information about how a people lived and died.*

Below: *The Nelson Collection contains a matchless array of Alaskan Eskimo artifacts from tools as common as our paring knives to objets d'art as cunning as model trains for adult hobbyists. The stone tools are scrapers and* ulus, *curved knives which would have had wooden handles. A Russian trade bead adorns one stone-bladed skin scraper. The long knife in the center has grooves for a surer grip. Fish scales still stick to the handle.*

Opposite: *The model kayak was made by a Norton Sound hunter as a miniature of his real, seagoing craft. Covered in sealskin, it is equipped with a lance, harpoons, spare paddles and a float board, a sort of buoy attached to a projectile and used to retard the escape of a prey that had been struck. Several full-sized hunting tools surround the model: carved seal darts (upper right), a walrus harpoon head (lower left), a smaller toggling harpoon for seals (upper left) and an ice scratcher (lower right). Nelson related that Eskimos believed finely made tools improved a hunter's chances because the spirit of his quarry would welcome capture by him. The ice scratcher, armed with real seal claws, was used by a hunter to scratch the ice in imitation of a noise that seals make. Thus it aurally camouflaged the hunter, made him invisible in a way and provided powerful magic by augmenting his own stalking skills.*

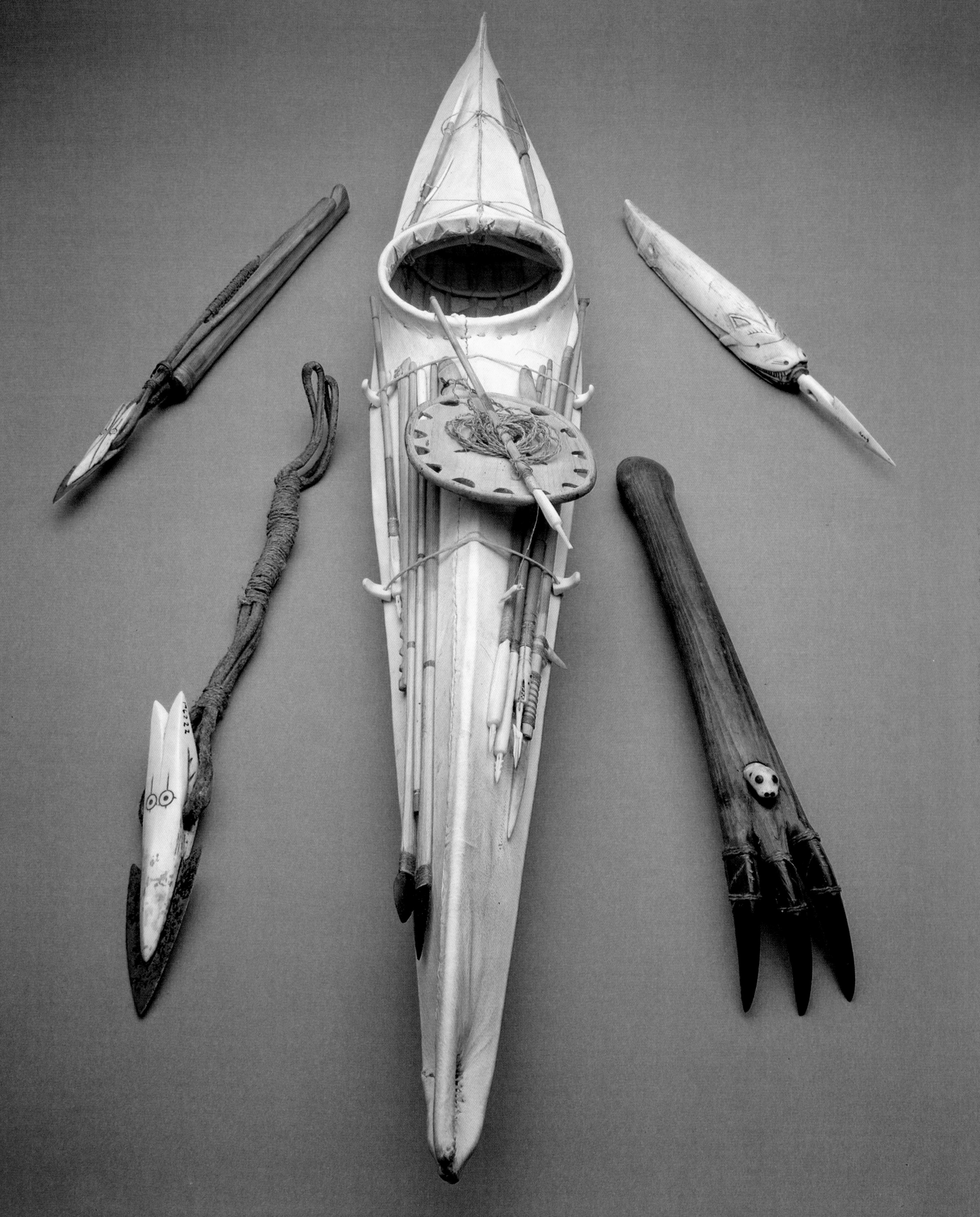

Opposite: *The wooden food bowl features a mythical beast with a central "life line." Trade beads replace the ivory plugs that would have been used before any European contact. Faces on the handles represent man-like beasts wearing masks to hide their identities. The ladles bear similar symbols, all of them painted with a seal-blood mixture containing charcoal for black, ocher for red.*

Left: *Shaped like a seal, the wooden box has an* inua*—the face of the seal's spirit—on its top and the animal's physical features below. Objects like these were designed to be viewed from every angle, also with utility in mind. Gently rounded, this box would stow easily in a skin bag or parka and its lid still fits so snugly that it requires no fasteners. The vessel might have held tobacco, which Alaskan Eskimos received in trade from Siberians, who obtained it through a chain ultimately leading back to eastern North American Indians in the sixteenth century. Thus long before Nelson's time the weed had become universally popular, traveling the long way around the world from America's east coast to the west.*

stone animals, skinning knives, wooden bowls decorated with the perplexing animalistic image of Tonngut "the keeper of the game," little gadgets used to link thongs together. As the Burgess Shale and Springer artifacts represent ancient phyla, these several thousand objects represent "a fossilized stage of humanity"—random physical relics of a culture that thrived in one of the earth's most inhospitable regions.

The most complete collection in the world of pristine Eskimo material, it reveals how closely these people lived with nature; how completely they used its resources, notably the birds, fishes and mammals that shared that place. (They not only made parkas from birdskins, but created clothing from fish and animal intestines.) Most remarkable, perhaps, is the adornment of the objects. Not satisfied with plain utility, these nominally primitive people decorated everything. The simplest bone tool displays

the image of a walrus or gull. Many an implement expresses visual puns: turned one way it looks like a raging polar bear, inverted it becomes a baby seal. Fishes become birds; biota lie in the eyes of the beholder.

What purpose can this collection have? It is a time machine. It can tell us, if we have the wit to learn, how these Eskimos lived and what they believed. It can reveal how some of our own humankind made life worth living. Beyond that, it contains beautiful things: inspiring creations, the embodiment of artistic talent which speaks to all mankind. Like the biological and mineral collections, this one preserves information about the past on earth—the human past.

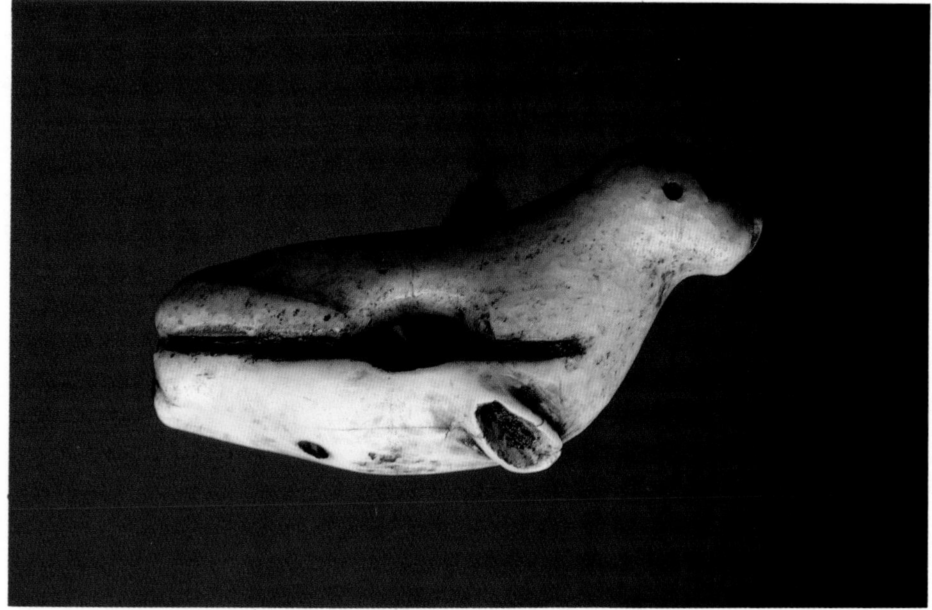

Opposite: *The Nelson Collection also includes these "line fasteners"—carved ivory buttons used to bind thongs together for any purpose from securing a bundle to making a chain of flexible leather links. In the uppermost one, its face a masked predator, one thong was laced through the top holes and another thong looped around the animal's neck. Tension on one thong held the other fast, yet the two could be easily disengaged even when frozen. (A knot, which could not, might easily spell death in the Arctic.) The smiling human faces represent men, the frowning ones women, according to the Eskimos' iconography. Lines through the eyes (top and lower right) represent concealing masks.*

Left: *The Collection also contains visual puns, as in a belt fastener depicting both a seal and a polar bear. In this case the implement represents both a common quarry and another animal known to hunt it successfully, so the tool serves twin symbolic functions in terms of addressing animal spirits.*

THE MUSEUM OF CONSERVATION

A word with many meanings in the world at large, in the Museum "conservation" implies a multitude of problems from "Peruvian pot disease" to kleptomania and a like number of preventive measures or delaying tactics. It means saving stored objects from accidental breakage and natural disintegration; protecting the exhibits from venal or capricious theft; preserving for posterity the raw information contained in a base rock, bird's egg or broken pot; keeping healthy the people who work here. "Conservation"—arguably any museum's single most important function —goes on in manifold ways:

- Collections—aggregations of potsherds, pterodactyls, live orchids, dead languages recorded on metal disks, innumerable objects of known or unassessable value—have unpredictable worth for scientific and scholarly posterity. Their safekeeping and accessible arrangement amount to de facto conservation.
- Security, in particular the protection of intrinsically valuable objects on display, assures the literal conserving of these things for posterity. The sophisticated fire and burglar alarm systems, the cunning design of exhibit cases, the deployment of uniformed guards and plainclothes officers, all protect priceless property held in trust.
- Environmental conservation as in "Save the Whales" or "Woodsman Save That Tree" are nagging concerns for Museum scientists. For the biological investigator, a living species' extermination, or the inestimably larger tragedy of an entire habitat's ruin, represents the irrevocable loss of pieces in the puzzle of life on earth.
- Personnel safety—the conservation of scientists and technicians —involves addressing a thousand threats, from muggers in Belize to deadly monkey virus in Tanzania and toxic chemicals back in the Museum's labs.

Caring for the collections, a variegated business, consumes the largest part of each department's energies. In Botany and Entomology small beasts must be kept at bay, many of them the same insects that housewives fight. In Anthropology a swarm of arthropods can certainly destroy a case of birdskin Eskimo parkas, but chemistry assaults these collections more often than bugs. In turn the Anthropology Conservation Laboratory

Ancient Egyptians mummified not only priests and royalty but embalmed many animals to accompany honored human dead to an afterlife in the spirit world. It now falls on conservators to preserve these artifacts in the more hostile environment of a temperate climate where fluctuations of temperature and humidity wreak havoc with organic material. A linen-wrapped cat mummy like this may require subtle care since animal material putrefies in moist air while the plant fibers of wrappings split and crumble if allowed to dry out excessively.

Every kind of material is susceptible to different agents of disintegration whether animal borers and moths, vegetable fungi or physical and chemical changes. The Tlingit box from Alaska, used to hold food a century ago, was carved in the shape of a bird from a single block of wood and must be protected from dry rot. The blouse bearing a "mola" design, made by San Blas Cuna Indians of modern Panama, is stored away from light to preserve its colors.

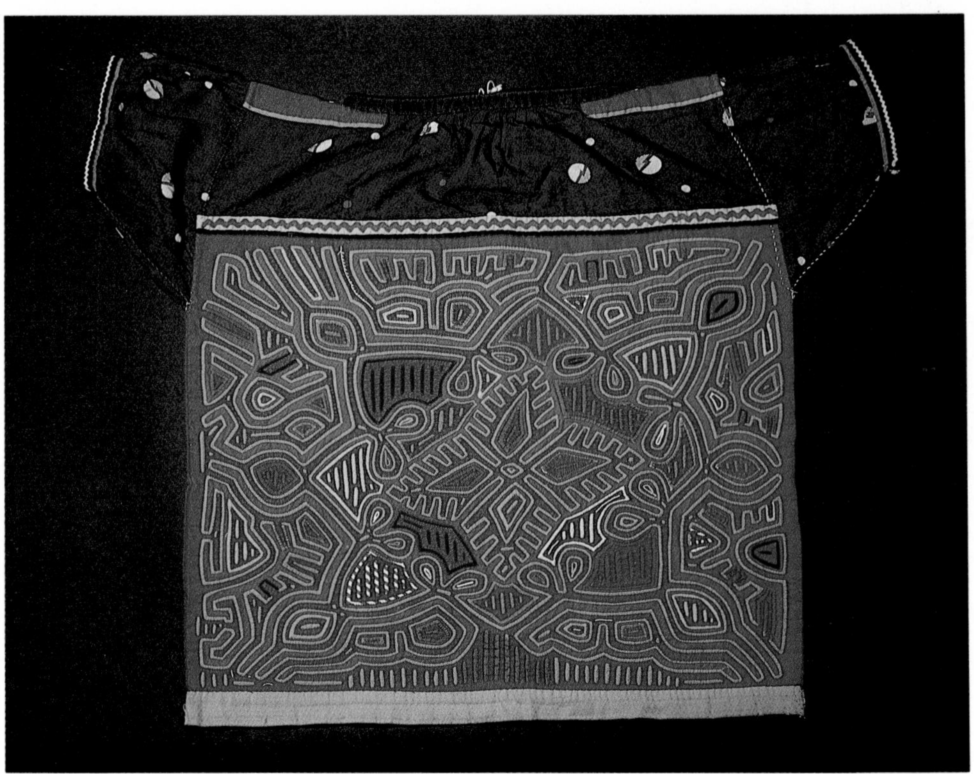

Right: *Conservators must protect this painted deerskin outer dress, made by Labrador Eskimos, from insects.*

Ultramodern analytical instruments punctuate the panoramic array of objects of the Anthropology Conservation Lab, which prepares artifacts for loan or exhibition. Near the century-old model umiak, built by an Eskimo who sailed such craft, lie a collapsed bag made of fishskin and dancer's "mittens" adorned with the beaks of sea parrots. The boat's bow points toward an Eskimo mask and a bronze figure of the Egyptian falcon god Horus wrought about 1000 B.C.

Archaeologists depend heavily on pottery for clues to prehistoric events. These fragments belong to the Valdivia complex on the coast of Ecuador, which begins about 3000 B.C. Numerous complicated techniques of decoration were employed, all of which are also represented on older pottery found on the Japanese island of Kyushu. The suggestion that the similarities result from trans-Pacific contact remains controversial although more than 15 years have elapsed since it was proposed.

Right: *This prehistoric copper figure, from the Etowah Mounds in Georgia, has been placed in a Plexiglas sandwich to protect the metal from environmental pollutants and to prevent damage when the object is handled by technicians or studied by scholars. Made of metal which Great Lakes Indians mined and used for trade goods, it represents a ceremonial dancer. The man's mace and peregrine falcon costume denote his high rank.*

employs chemistry to ameliorate problems brought on by time or neglect. Artifacts often disintegrate because of substances within them or in their innocent surroundings. The cellulose of a wooden drawer or wicker handle produces acids that harm pottery glazes. Moisture can leach salts out of old clay, or the combination of earth elements and decorative pigments can make a surface part company with the object itself.

The Lab's first rule might have been taken straight from the Hippocratic Oath: First, Do No Harm. This represents a change of curatorial mind, says the facility's head, Carolyn Rose. Previous museum people took it upon themselves to make objects look spanking new, even to "improve" them according to their own criteria. Lack of technological sophistication and devilishly good intentions often destroyed ethnologically significant information which an object contained. Restorers of the old school often made a specimen into what they thought it should be, and neglected what it actually was. The newer, sounder approach settles instead for preservation. Thus if a vial contains an ugly discoloration, the conservator removes only enough to analyze, to learn what the vessel once contained—oil or perfume—and provide new information about its use. Analysis of pollen grains in a basket's weave can reveal what crops its makers knew. Even anonymous labels are saved for potential study since a future curator may recognize the antique scrawl of the collector and pinpoint more about its origins.

The work requires unusual sensitivity to both materials and cultures since it involves chemistry and anthropology in about equal parts. While people of the tundra coasts used everything from the seal except its bark, different animal tissues respond differently to different curative materials, just as waxy plant fibers can be ruined by the very solvent that can save a waxless mat from rot. A broken feather on an Indian headdress might be the result of careless storage or the symbol of some deed its owner performed, so it is not automatically replaced. Corrosion on a metal object might dissolve in a dozen solutions, but perhaps ten of them will change the chemistry of the surface alloy, often the key to discovering where the object came from. (Thus the encrustations must be removed "mechanically," scale by scale with tiny picks.) In many cases the conservators decide it's

best to do nothing at all because, for example, the flaking pigment on an Mbe drum from Gabon can't be consolidated without changing its color.

When actual restoration work seems appropriate, Mrs. Rose says, the conservators abide by the Rule of Reversibility and the Six-Foot/Six-Inch Visibility Ethic. The first states that anything done must be potentially correctable. If a pot's missing handle is to be replaced, the new one must be glued on with an adhesive that's soluble in a substance that won't harm the rest of the object. Then it can be removed easily if necessary. (Further, vegetable glues must contain a fungicide and insecticide so that they won't support micro- or macro-organisms.) The second rule states that good restoration should be invisible from six feet away; but from six inches it must be obvious where the original material ends and the reconstruction begins.

Every object that passes through here is carefully researched and all the work performed on it recorded for future reference. Most of the specimens are bound for exhibition or loan. Thus the Lab packs each object, whether for a trip across the Mall or halfway around the world, with extraordinary care and odd skill. On a single day the Lab's three staffers—and several George Washington University students earning academic credit as interns—might be working on all of the following objects: a bronze statue of the Egyptian falcon god Horus from the twenty-sixth Dynasty; a brittle Eskimo parka made of salmon skin with caribou trim; a small arsenal of wood-and-bone bird spears; a fossil bone chipped for use as a tool; a collection of European pigments used by American Indians; an Etruscan urn; a many-tiered African crown made of wood pith; a sword of Damascus steel; a Tibetan saddle made of leather, iron, wood and the skin of the giant putative "snow toad." (This animal, unknown to the herpetologists downstairs, might yet prove to be the problematical Yeti's prey—or predator.)

Anthropology also copes with the unusual curatorial problem of nonpreservation. Among the ethnographically priceless objects collected in the last century are religious artifacts from a variety of cultures for whom artificial preservation seems anathema. Many masks and talismans were wrought and mystically endowed in the knowledge that they would

Opposite: *African musical instruments, whether wood-and-hide drum, shell-encrusted thumb piano, painted-gourd rattle or reptile-skinned arched harp, challenge conservators with distinct combinations of preservation problems. Tanned leather contains oils which affect metal and wood. A wooden zither may contain destructive insects. When made and used, many objects were daubed with earth pigments, dairy products, even dung—materials which themselves deteriorate and erode other substances. The question then arises whether to remove the destructive material or preserve it as culturally significant.*

The Tlingit raven rattle provides a dramatic display of complex Northwest Coast Indian iconography. The rattle as a whole represents a raven—with the head at the outer end and wings at each side of the handle. On one side of the body is a hawk's head, grasping the tongue of a recumbant figure that possibly represents a human or a humanoid frog, since frogs often appear in this position on other rattles of this type which allude to a myth.

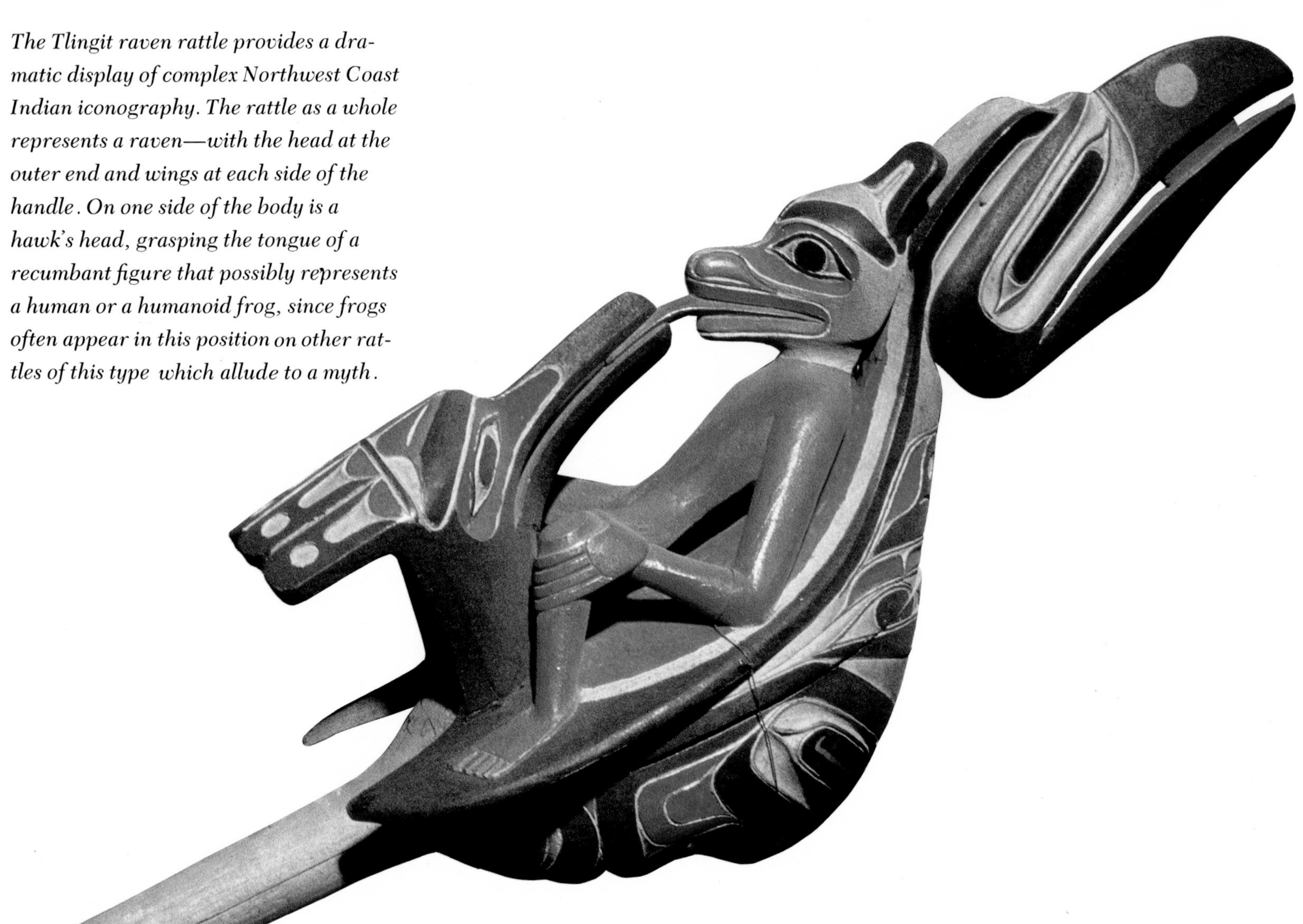

Opposite: *Exhibits people try to display objects in natural positions while conservators hope to protect them perfectly. These aims can conflict—as with an object like this Plains Indian war shield. If displayed for long, its feathers could drop off and its frame sag; if perfectly preserved it would lie horizontally in a dark drawer and not be seen at all. Consequently a kind of compromise exists. Exhibit cases are painted with acid-free paint (to inhibit acidic wood vapors), mounts for each object are designed to distribute weight evenly, displays are designed so that objects of different materials don't touch and lights are specially filtered.*

eventually disintegrate. A widespread belief—witness the restless yuletide ghosts in Dickens' *A Christmas Carol*—holds that under certain circumstances living spirits can be held captive in one kind of limbo or another. Thus some religions hold that a numinous object should be allowed to decay naturally over time and release the spirit it possessed (or which possessed it). This, of course, runs contrary to any museum's basic purpose of preserving objects which may be absolutely irreplaceable, given that the culture which produced them died out. Even latter-day tampering runs counter to conventionally good conservation, yet some surviving Amerindian tribes find themselves honorbound to renew their ancestors' medicine bundles. Old tobacco and grain by tradition must be replaced with fresh; brittle leather thongs that bound the bundles together need overhauling.

One tribe in particular expressed deep concern about its ancestral legacy here, and its elders visited Washington. They made no claim that the artifacts had been acquired by underhanded means and did not demand them back. Seeing some things moldering in one attic, a shaman told Mrs. Rose, "Don't change a thing. They're disintegrating beautifully." They did have several requests which the Museum ultimately granted. The elders identified objects of special religious significance, which were to be set

Preceding pages: *The pelts of 6,000 large animals—wolves, bears, seals, sea otters and the like—hang in the Fur Vault's cold dry air. Theodore Roosevelt shot three of the African lions preserved here and one of the bison skins came from Buffalo Bill's Wild West Show.*

Opposite: *The first rule for conserving animal material, whether bug, bird or bison: keep the vermin away, usually by dosing collections cases with powerful insecticides. Even then, imperfectly prepared skins of some birds will self-destruct as fatty residues oxidize and cause the specimen to disintegrate. This colorful array of admirably preserved tropical forest birds includes: nine fairy bluebirds from Southeast Asia, a pair of red-bearded bee-eaters from Siam (lower center), a pair of carmine bee-eaters from Africa (lower right), a blue-crowned motmot from Central America (upper right) and two New World trogons.*

apart from the profane ones and kept separately. They are never to be publicly displayed, nor may lay members of the tribe see them privately lest the captive spirits harm either those individuals or the entire tribal nation. So be it. If the Museum acquired an object honestly (according to the mores of the day) its administrators do not feel compelled to surrender it to tribal heirs. However, the Museum firmly intends to treat artifacts of any sacred nature with every respect its original ethos requires.

(Speaking of preservation and anthropology, Museum people themselves represent some curious phenomena which illuminate how slim the differences really are between supposedly disparate cultures. Consider the practice we abhor among many so-called "primitive" people who sacrifice a puny twin at birth so that the stronger sibling might surely survive. One departmental chairman knows of an 1880s case in which such a runt was placed on a windowsill and left to die until an aunt took pity on the little thing. He grew up to become a legendary zoologist—albeit a hypochondriac—who finally retired and was last heard from lucidly enjoying life at 94. His robust twin, as it happened, died in childhood. Who were the "primitive" parents? Upstanding Maine Yankees.)

In the Vertebrate Zoology Department, preservation and conservation are practiced selectively in a sense. Some specimens are entirely preserved; for obvious reasons the soft parts of others are abandoned and only the skin and skull (or entire skeleton) saved for study.

Taxidermy, the traditional method of life-like preservation of vertebrates for display, was practiced extensively here until a generation ago. While cosmetically appealing, the mounting of a specimen is a difficult, expensive and labor-intensive task. The one remaining technician with these skills uses them primarily to maintain and repair old specimens such as the Rotunda elephant whose hide must be vacuumed and lubricated with beeswax every few years. (To mount it in the first place took the better part of two years.)

One shortcoming of taxidermy lies in the fact that it ignores the skeleton; the part of a vertebrate's body most crucial in taxonomic studies. A skull or skeleton might be saved and prepared for study, but this has nothing to do with the "mounting." (For the record, "stuffing" an animal

Once threatened with extinction by wanton hunters, the walrus is now making a comeback in the cold-water wild. These sea mammals use their tusks for fighting and—in the manner that mountaineers use climbing axes—as tools to haul themselves out of the water onto ice floes. They may also use them to pry their food—clams and such—from sea-bottom beds. (A taxidermist preserved this bull's massive dignity.)

lies beneath the taxidermist's art, though it has its place in museum work. Bird and rodent skins are stuffed with cotton wadding, a swift and simple process, so that the pelts will keep their shape and be easy to handle.) By comparison, skeletal preparation is a peculiarly esoteric business. As the head of the Osteology Preparation Laboratory puts it, "The technician hasn't been born yet who can work as carefully as the bugs." Douglas John's job centers around the care and feeding of generations of perversely useful insects. He may be the only parascientist here with a chronic case of beetlemania.

These bugs, formally designated *Dermestes maculatus* and *D. ater*,

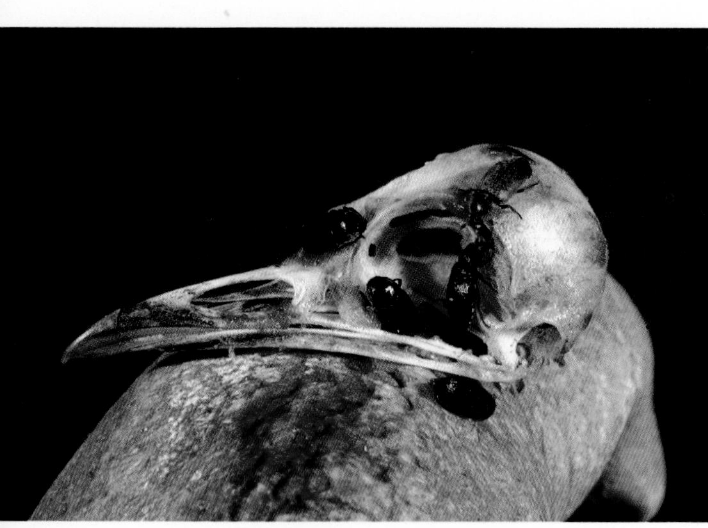

Dermestid beetles at work, cleaning a bird skull.

have been famous carnivores in museum circles for two centuries. In fact a member of the genus next door, *Arthrenus museorum*, goes by the familiar (or overfamiliar) name "museum beetle." When Linneaus himself discovered and described it, the insect was infesting a museum, devouring valuable collections of zoological specimens. Here and now, however, dermestids are encouraged to consume specimens under ideal conditions within a pair of walk-in freezer chambers which have been adapted to keep things warm and moist instead of cold. The temperature hovers around 80° Fahrenheit and the humidity a damp 65 percent. It is made even nicer by the addition of plenty of cotton batting for the small beasts. The cotton, serving the same purpose as fur or feathers in the necrological wild, offers adult beetles a place to lay their eggs, for the eggs to hatch out caterpillar-like larvae, for the larvae to bore holes and pupate into adults and for them to lay eggs again during an 18-week life cycle. The icing on this environmental cake, as it were, lies in the wealth of food available—an endless supply of meat.

Eating meat is how the assiduous bugs make their mark, indeed how they earn their keep. The Museum doesn't go to all this trouble and expense just to nurture a colony of prolific insects naturally found all over the world. The dermestids do important work. They clean fresh bird, mammal and fish specimens down to the bare bones, which is what, after all, an Osteology Preparation Lab prepares for the vertebrate zoologists upstairs: cleaned skeletons for taxonomic study.

Dermestids are uniquely suited to the task in captivity because both larval and adult forms practice exclusively carnivorous habits. They disturb nothing but animal matter except when preparing to pupate. (Then they'll bore into anything soft enough for a fingernail to scratch. This means their chamber's seams cannot be sealed with tin or lead solder because the larvae would gnaw their way out. Apropos, an old sailing ship nearly sank because beetles had infested its cargo of penguin carcasses and the pupae, preparing for metamorphosis at sea, honeycombed the wooden hull.)

Their overseer won't even guess at the number of his charges productively inhabiting the preparation chambers. There are enough to cover every specimen in the 600-cubic-foot freezer boxes. The speed of

Preceding pages: *Manic hunting in the name of perverse "sport" nearly made the bison extinct. The beast, which once ranged across North America, was saved by growing interest in the biological world in general and an early conservation campaign in particular. Both of these causes were advanced by the Museum, which houses this diorama.*

the dermestids' performance depends on the amount of work. A small mammal may require a week in the chamber, though a particularly fresh one may be ready overnight. A sea turtle, which has very complex musculature within the shell, needs as long as a year. To hasten the work, technicians remove accessible flesh by cutting it with scissors and knives. This process might be likened to Athenian slaves roughing out stones before Phidias fashioned them into the Parthenon's Elgin marbles. Any fool can do the bull work, but it takes special ability to finish it off. While a human might wield a scalpel with surgical precision, only a beetle can get into every tiny, convoluted space within a bird's spinal column or the sinuses of a reptile skull.

In the course of a year as many as 2,000 specimens pass through the damp, dark, warm chambers and come out clean white bones. When ready, each specimen is removed, bugs and all, to a "cryofumigation" chamber, actually a real freezer. Deep cold dispatches the bugs faster and more certainly than any insecticide; also more safely as far as the human attendants are concerned. While a chemical requires three days before people can safely open the killing chamber, a freezer euthanizes the bugs overnight with no harm done to the cleaned skeletons and no residual danger to anyone.

Problems remain, of course. For one, quantities of gasoline are used to remove grease from old specimens, then a series of ammonia baths removes the gasoline, all of which means working with hazardous and caustic substances. For another, the Lab crew was grumpily accused of sloppy work by some ornithologists who said the birds they sent down in toto came back jumbled up. One numbered skeleton would be missing a tiny circular eye bone while another would have one too many. (It turned out that the technicians weren't to blame, the beetles were. The bone would get snagged on a beetle's hair and toted between compartments in the specimen trays. When this transport was discovered, the trays were re-designed.) The worst threat lies in the occasional escape of a few pregnant denizens of the Osteo Prep Lab, located in a courtyard, to the surrounding building. They threaten not only zoological material inside; dermestids—commonly known as hide beetles—could bring down the Arapaho teepee

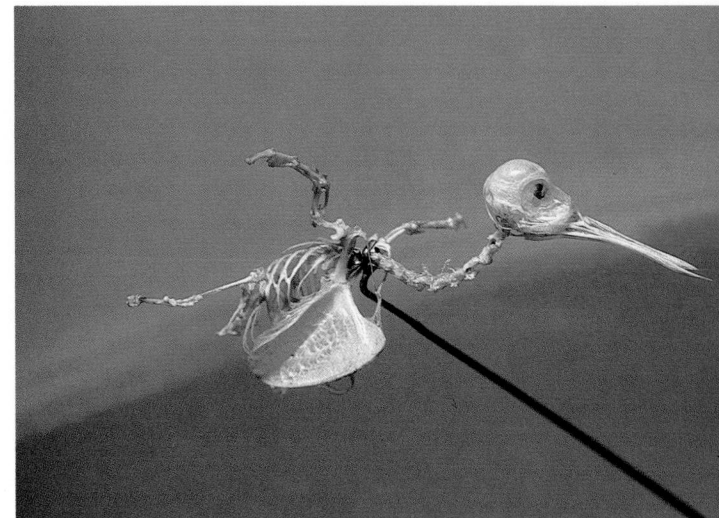

In the wrong places, the carnivorous dermestid beetle can cause inestimable destruction. But it earns its keep in one Museum precinct—the Osteo Prep Lab—by cleaning specimens down to the bone. Neither the hummingbird skeleton (above) *nor Russell's viper* (overleaf) *could have been prepared for study as well by human hands alone.*

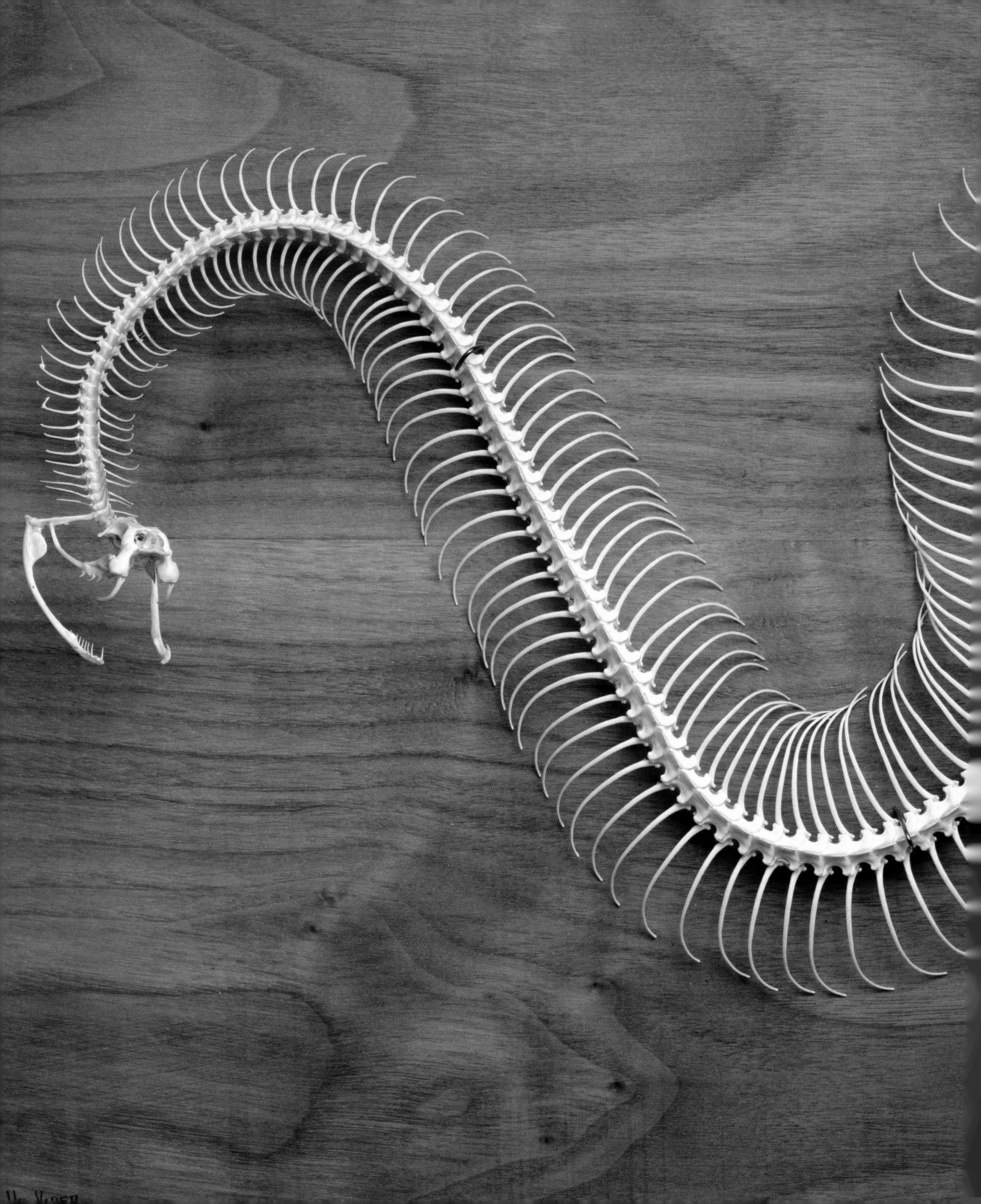

in the North American Peoples Hall. Several years ago some beetles got into a reptile diorama and devoured a small snake. Before anyone noticed, only the coiled skeleton remained of the freeze-dried specimen.

But for the fact that it leaves too much nutritional matter available for stray bugs, freeze-dry preservation of specimens for exhibit might replace other methods. It simply circumvents decay by a remarkable process: bacteria, enzymes and other bioactive minutiae require a water medium and cannot survive without one. This basic fact has been empirically applied since people hung the first meat in the sun to form jerky or shriveled grapes into the first raisins. The removal of water, whether by airing out in the arid Andes or baking in a crisper oven, preserves the remaining material by rendering it unfit for bacterial habitation. What works for foodstuffs applies to museum material as well—or would but for one snag. Everyone expects wrinkles on a prune, but a bird that shrivels up has little use in a biology collection and less in an exhibit hall. Drying might preserve the specimen but under ordinary conditions spoils its appearance by drastically deforming it.

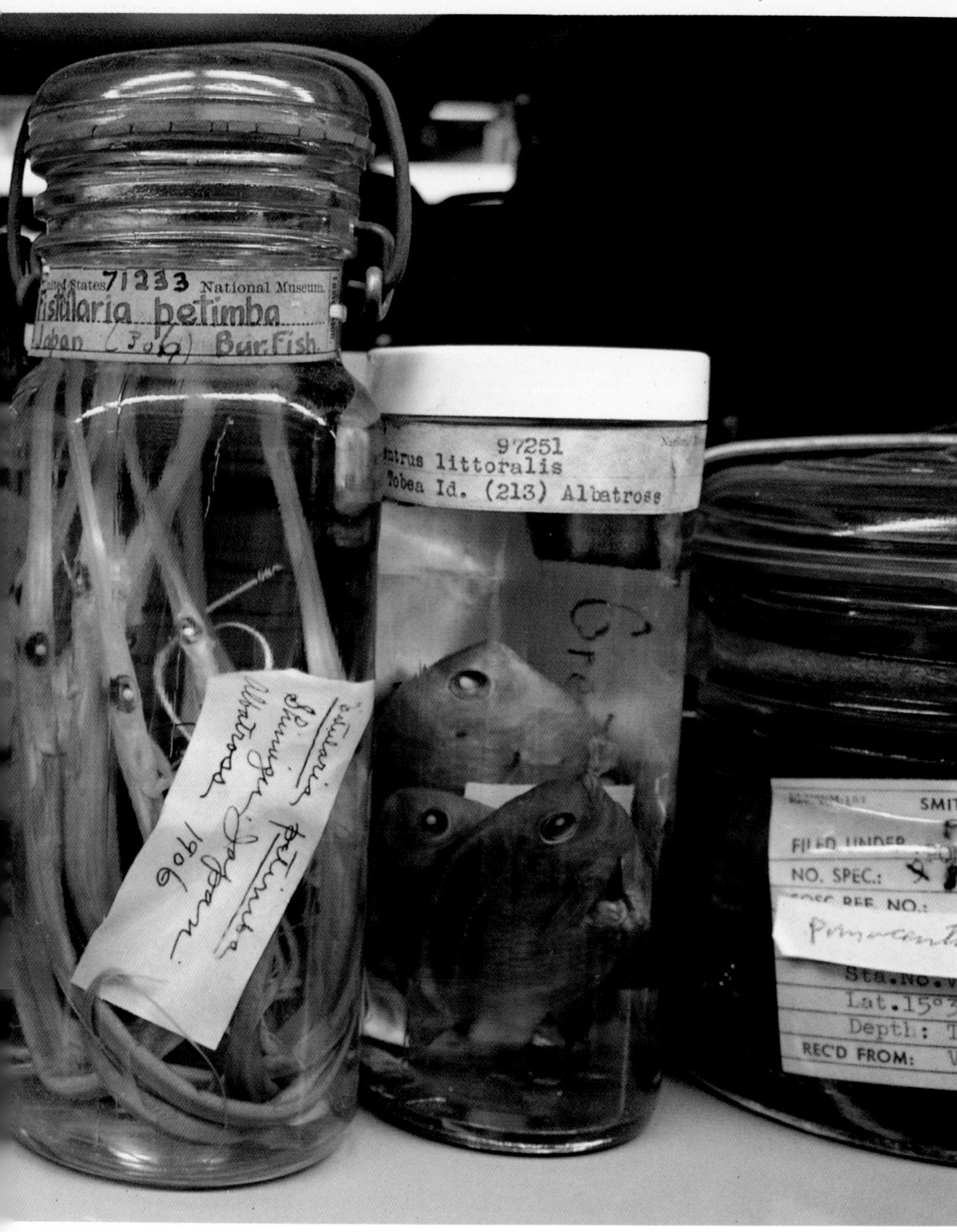

To preserve fishes whole, ichthyologists store them in alcohol or formaldehyde solutions. The fluids require special handling since one burns readily and the other has toxic properties.

The culprit here is surface tension. As water evaporates, the remaining fluid forcefully pulls together whatever contains it. This means that the fibers of a wool sweater get compressed and tangled. It means the walls of once-living cells break as they surrender the fluid that allowed their vitality. It means the shapes and appearances change radically for the worst. The answer then lies in evading the harsh effects of surface tension by avoiding the liquid state of H_2O and, instead, dealing with it as a solid. Just as water evaporates, ice "sublimes." This is what has occurred when you open the refrigerator after a vacation trip and find all the ice cubes have shrunk.

The secret for the Museum's purposes lies in manipulating two controlling factors, temperature and pressure. All that involves is some heavy-duty gear that capitalizes on the elementary laws of physical chemistry, laws which are applied with chilling effect in the Freeze Dry Laboratory. At sea level, where barometric (that is, atmospheric) pressure hovers around 760mm of mercury, water boils at 100° Centigrade. At an altitude of 5,000 feet, where the atmospheric pressure is less, water boils a

A bird, mounted in a lifelike pose, joins the menagerie within the freeze-drier. When sealed, the machine uses low temperature and high vacuum to remove all moisture in the form of ice crystals. Plants and animals preserved this way don't undergo ordinary decay because the micro-organisms which cause rot and disintegration require wet habitats. But most specimens bound for long-term exhibition are still preserved by standard taxidermy because, just as people enjoy freeze-dried coffee, carnivorous insects consume freeze-dried animals whose tissues remain as palatable as fresh ones.

Opposite: *A freeze-dried eastern screech owl demonstrates the lifelike results of this new technique.*

Following pages: *Napoleon I gave his consort Empress Marie Louise this diadem in 1811. Set in silver, its 950 diamonds weigh 700 carats. Its original emeralds have been replaced with 79 cabochons—polished but unfaceted gems—of Persian turquoise.*

few degrees cooler. Atop Mt. McKinley, where atmospheric pressure is still lower, water boils at such a low temperature that a mountaineer can't boil potatoes. The low pressure reduces the vapor point of water below culinary utility. Applying a corollary to this rule, at room temperature (theoretically 26° C) water will boil if the atmospheric pressure drops to 25mm of mercury or about 3 percent of the sea level norm. Needless to say, this requires a special room, either one built into a spacecraft or a sealed terrestrial environment fitted with a powerful vacuum pump. The latter was chosen in 1959, when the museum world's first freeze-dry lab was built. Hulking in a basement, it looks something like an old apartment-house boiler. The device contains two connected chambers, each of them equipped with a refrigerator coil. (Scientifically these are considered "heat sources" since they provide thermal energy on the Kelvin scale which starts at absolute zero, 420° below zero Fahrenheit or −273° C). The coil in the first chamber brings the temperature down to −20° C while the vacuum pump reduces pressure to 50mm of mercury, about the level encountered at an altitude of 40 miles. Under these conditions frozen water sublimes from the solid state of ice into frozen vapor, a gas. The cold vapor circulates into the second chamber and encounters a colder coil kept at −40° C. There it condenses like breath on a winter windowpane, effectively going out of circulation.

The beauty of these phenomena becomes clear when the first chamber is filled with specimens: birds, reptiles, small mammals, plants or whatever else the Exhibits people need next. In due time—two weeks for a robin-sized bird or ten months for an 8-foot komodo dragon—all the water in the specimens' tissue sublimes away. As vapor it diffuses through frozen cell walls molecule by molecule doing no structural damage en route since no hydraulic surface tension comes into play. Some proteins change in the process, but in form and appearance the entire animal remains. Except for water content, which made up perhaps 60 percent of a specimen's weight, the whole animal remains. It only lacks the protoplasmic water which bacteria need to go about their decomposing business.

Even after the animal is returned to normal environments, it is in effect frozen. Each cell wall has become rigid since the cell itself lost the

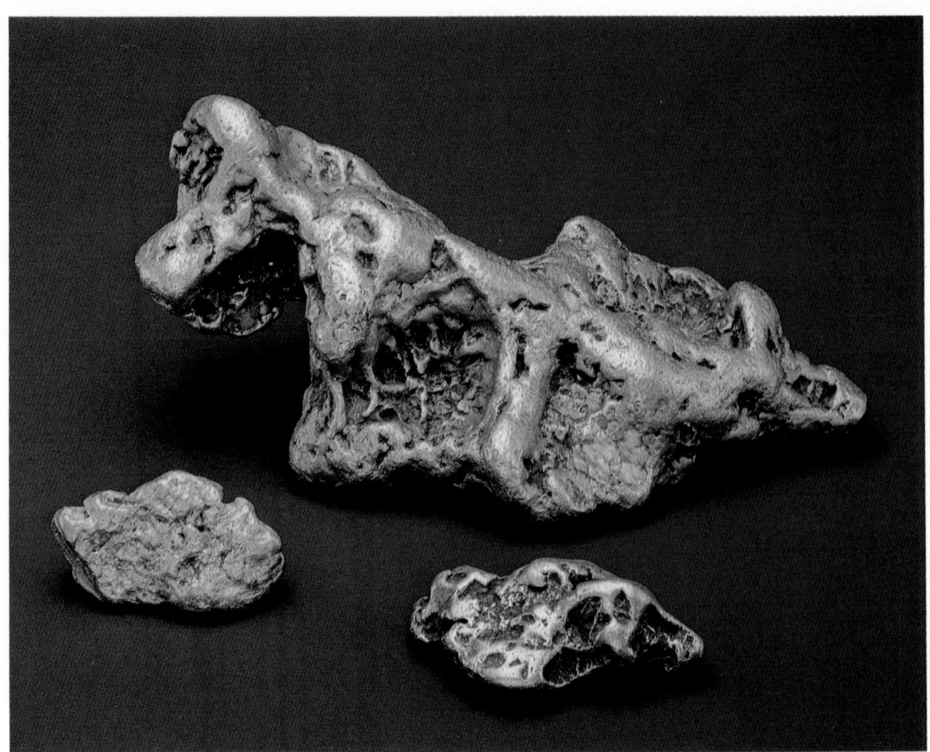

Specimens like this jewelry (opposite) *and the gold nuggets* (above) *pose special security problems because of their commercial value. Napoleon gave the diamond necklace to Empress Marie-Louise upon the birth of their son. The earrings were owned by another French consort, Marie-Antoinette. The largest nugget, which commands interest because of its great size, weighs more than five pounds.*

water that allowed it to be flexible in life. The animal holds the pose it was given before entering the chamber. Birds, mammals and reptiles hold their original colors too, though insects, fishes, amphibeans, and crustacea fade so their colors must be restored with paint. As for preparing a specimen, all this requires is placing it in a chosen pose and supplying enough support to hold it steady. This involves far less human work than taxidermy; a freeze-dry technician can prepare five times as many specimens in a day as a skilled taxidermist. Specimens are prepared as they arrive, then kept refrigerated until the biweekly opening of the machine. Specimens that have fully sublimed their moisture are removed and new ones added as space permits. The chamber is resealed again and within 90 minutes the vacuum and supercool temperature restored.

Freeze-dry technology has applications in some unexpected realms. Bog-soaked fossil bones, too wet for ordinary conservation, have surrendered their moisture and remained intact. Water-damaged documents, including the log from a ship sunk a century earlier, have been saved in the freeze-dryer. Food for backpackers, field scientists and armies alike has been made this way, while cryotechnology enables the longterm storage of drugs, antivenin, even blood without refrigeration. The food industry doesn't need to lower temperatures much below the freezing point; bacteria-supporting water can be removed and the shrinkage that occurs doesn't harm the product. Medical cryobiologists have higher standards, however, as they preserve uncontaminated and undeformed bone, nerve and skin tissues for human transplants. In time freeze-drying may become a prime technique for preparation of Museum exhibits—if a means could be found to discourage vermin from using the beautiful specimens as food.

Physical security will remain a problem in every museum so long as

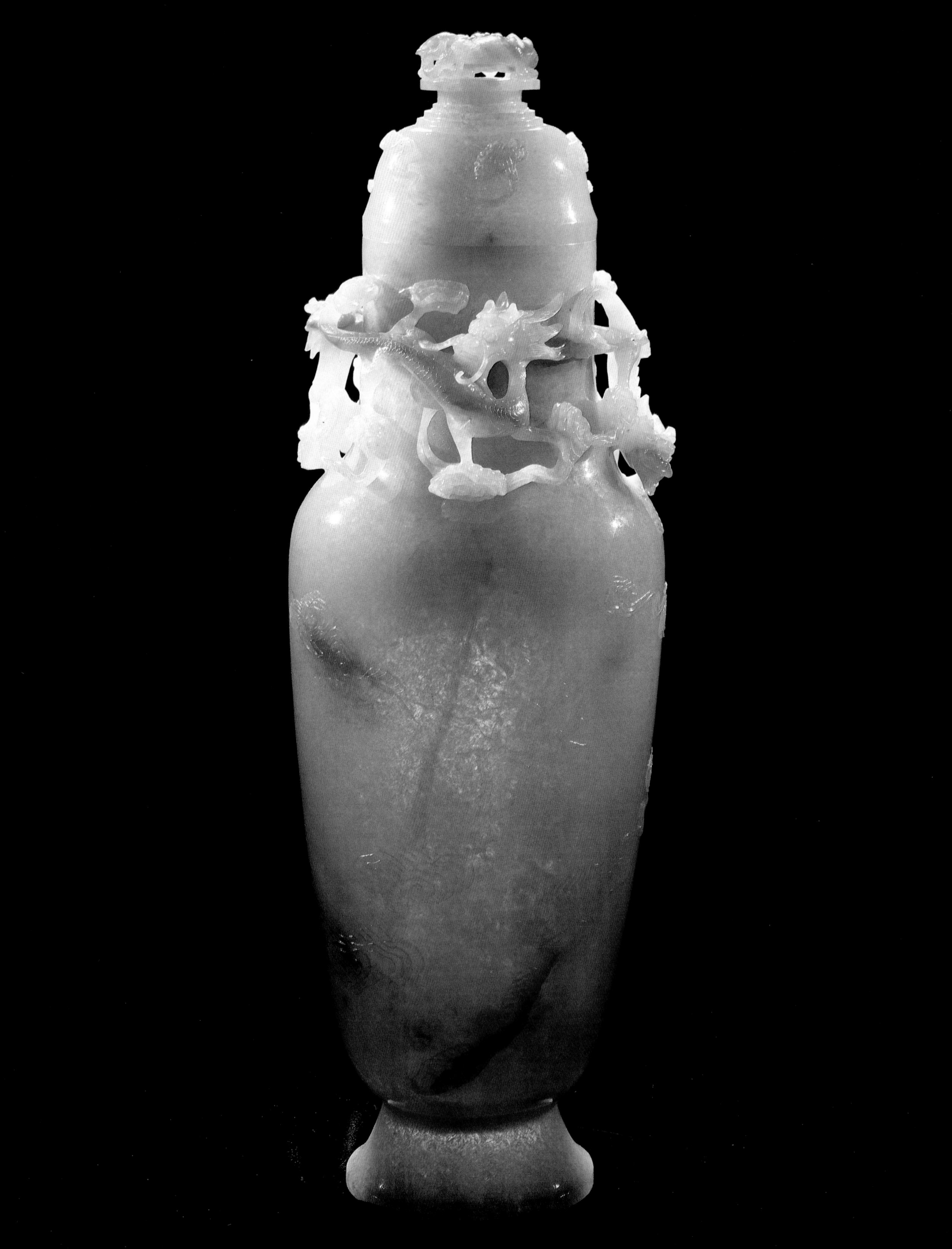

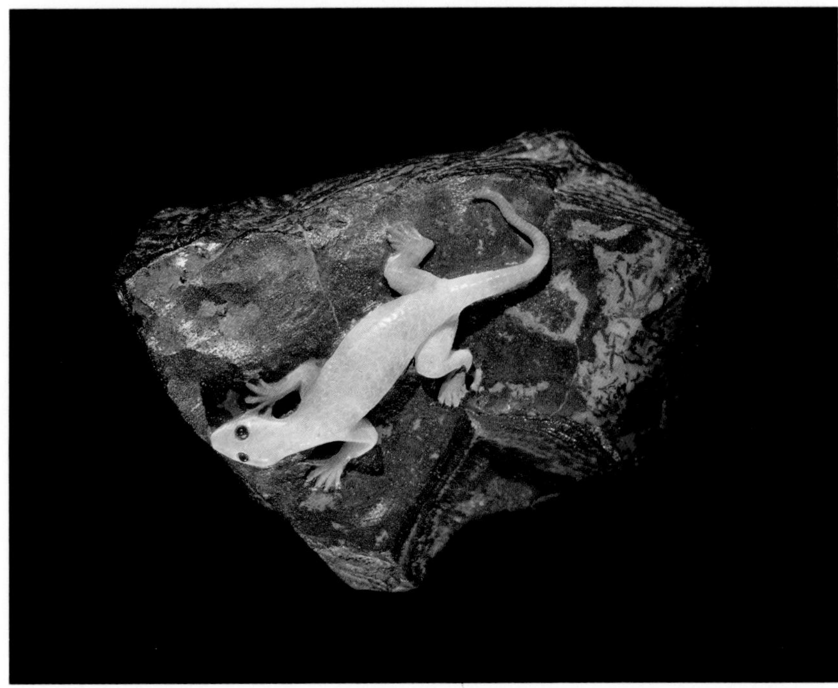

Opposite and above: *The intrinsic value of many Natural History objects (and thus concern about their literal conservation) resides in the combination of a rare raw material and unique artistic embellishment. Each of these three lapidarian treasures—an oriental jade urn, a labradorite medallion and a miniature Australian opal lizard—was wrought from a single stone.*

rare objects have financial or esthetic value. Time was that fossils were coveted only by scientists and a few esthetes who recognized their curious beauty. Now these biological relics have found a commercial market at a time when people collect everything from beer cans to antique greeting cards. Before the fad became intense, the Museum took what were considered appropriate steps to keep the remains of vanished life from vanishing. Then one primitive angiosperm, not only a type specimen but the only known example of such an early flowering plant, disappeared from an exhibit case. Fortunately it had already been studied at length and described scientifically; all that contemporary techniques could reveal had been recorded. But no one dares say what future investigative method could unlock more knowledge or what significance it might have when compared with fossils yet unfound. Who would have taken such a thing from a locked case and what might have become of it? If it were stolen for a lark, it might have ended up in a trashcan. The other possibilities warrant no more detailed description than—for obvious reasons—the new precautions to safeguard such things.

Some years ago a fossil skull of considerable scientific importance vanished from a collection and in short order FBI agents interviewed everyone who had access to it, including a young man doing postdoctoral work in the Paleobiology Department. Checking further, investigators found the skull on his mother's coffee table some states away. It was retrieved undamaged, but then the doting parent complained that the agents had ruined her boy's promising career in paleontology.

Needless to say, some objects are guarded more carefully than others: the Hope Diamond, the "Baby Doll" reptile, the elephant tusk collection. On the other hand some specimens are even touchable: the *Brachiosaurus* thigh bone, the cast of a *Tyrannossaurus rex*'s head in the new Dinosaur

The continued preservation of human remains on exhibit, such as these mummies, involves two relatively simple procedures: maintaining an effective insecticide within the case to deter pests, and keeping a constant supply of moisture-absorbing chemicals in containers underneath the mummies, to ensure a dry environment. The prehistoric Peruvian mummy, buried in an upright, cross-legged position, was preserved by natural processes in an arid desert grave. The Egyptian woman, lying on her back, was artificially preserved by the ancient practice of embalming.

Hall, and Herbert Read's monumental bronze sculptures around the upper Rotunda whose toes are kept polished bright by passing visitors' hands. But the great majority of things go on exhibit out of reach or in cabinets which are at least securely locked and most often fitted with alarms. Even a cabinet that has no visible lock cannot be opened on whim; screws securing the lid don't fit ordinary screwdrivers and there are always some concealed fasteners. Objects that might have special appeal to thieves receive special attention from the guards, of course. The popular moon rocks, secure in cases which NASA approved, get jostled a lot. The bells in this room go off so often that the guards keep a pad of alarm forms in easy reach under a nearby rock specimen.

The guard force is understandably most proud of its response to a recent security breach that had nothing to do with objects of scientific value, just cold cash. A man conned his way into the Museum before visitors' hours and presented himself at the bookstore as clerks prepared for the day. He demanded what they were putting in the till—at gunpoint. The clerks complied and when he fled with the money they called Security. The building doors were sealed before he could get out and alert guards arrested him.

Freeze-drying can occur naturally. The head, leg parts and hide of this big-horned bison came from Alaska where Bison crassicornus *sank in a bog 28,000 years ago. The corpse froze, then dehydrated slowly until it was unearthed by goldminers working with high-pressure water hoses. Using materials at hand, the scientist who acquired it in the field sewed edges of torn skin together with white dental floss. The leg of a wild horse came from a similar site.* Equus przewalskii *now survives only in the Altai Mountains of Mongolia.*

Still, it happens that a guard might be overzealous. Among the myths that won't be laid to rest is the story of the new security man on the night shift. Making his rounds through the silent empty halls one winter night when the wind rattled the windows and rain lashed the roof, he got nervous. But true to duty he stayed on patrol. He heard a sound and challenged whoever was there. He saw a man's silhouette, heard another noise and called out again. The streetlight flickered through the trees outside a tall window. The shape failed to speak. A beam creaked, a mouse sneezed. The frightened guard drew his pistol and fired—plugging the mannikin in an anthropology hall. Or so the story goes.

Another tale I was unable to run to ground involves the loss of a type specimen, the corpse of an organism which was painstakingly described by the taxonomist who named it, the single specimen against which all others of its species are compared. It vanished after the administration began taking special pains to preserve the preservatives from on-the-sly drinkers with remarkable constitutions. Ethanol, a common lab fluid, is the essence of every good liquor, and some that aren't so good, from moonshine to betel nut beer. (Medical students have known this for generations as did every "wet" scientist who survived Prohibition: Ethanol can enhance a stuffed olive as well as the finest London gin.) Persons unknown—the guards were accused most often, perhaps unfairly—availed themselves and tippled from huge jars of the nearly pure stuff which rates about 199 proof. When the authorities ordered the supplies be kept under lock and key after working hours, most of the leakage stopped. But some drinker failed to mend his ways, so the story goes. Plenty of the precious fluid remained accessible around the Museum; it just happened to be in jars also containing specimens. This toper's habit might have gone undetected indefinitely because lab alcohol evaporates even in the tightest vessels. But he not only got in the sauce, he got into the type specimens and in the throes of legendary thirst swallowed one of the priceless organisms.

I heard the story from an entomologist who vowed it occurred in Mollusks. A senior malacologist, denying it happened in his realm, referred me to Crustacea. The dean of crabs couldn't remember his division ever losing a type specimen that way, but he knew the story too. "It happened

Overleaf: *Museum botanists maintain greenhouses at the Old Soldiers' Home in northwest Washington, D.C., to grow selected plants, study their living anatomy, then preserve specimens in the National Herbarium, the heart of the Botany Department collections.*

An orchid from Southeast Asia.

years ago," he said. "I heard it was in Fishes." Wherever, it represents an old tradition: Peter the Great bought barrels full of exotic specimens in the seventeenth century. Packed in rum, they never reached Russia because sailors drank all the preservative en route across the North Sea.

If the Museum has inevitably but sadly lost a few specimens to greed and one to dipsomania, it lost another to psychiatric aberration. A man entered the Reptile Hall, smashed the glass of a diorama, stepped inside and hacked away at the snakes with a machete. When the guards arrived and finally subdued him, several skins, most of them mounted on plaster forms, lay in pieces like sausages with their stuffings coming out. Asked to explain his unusual behavior, he said one of the snakes owed him $20,000.

The medical security of scientists and technicians is a far more delicate thing to assure than the safekeeping of any object. In the old days arsenic powder was used routinely to keep storage units free of vermin and museum people everywhere routinely must have suffered the inevitable results of long-term arsenic poisoning. Now a better understanding of various pesticides' pharmacology—and enforced Federal standards covering their use—reduces the risks. For decades, nonetheless, biologists watched some of their older colleagues succumb to cancer of the liver and pancreas, possibly a function of too frequent exposure to formaldehyde. Aware of the possible correlation, Museum people today assiduously turn on the ventilators when working with open jars of specimens. The fish and reptile labs in particular are all equipped with hoods and exhaust fans to carry off pernicious gases.

In the field, where some curators try to spend at least three months a year, less predictable problems arise in the hazards that all adventurers face. One entomologist, who decided as a boy to become a museum man when reading the William Beebe adventure book series that lionized a curator, keeps his rabies shots current. The reason: many insects inhabit caves, but so do hordes of frequently rabid bats. Primate zoologists also have to worry about diseases carried by monkeys, viruses which can easily infect phylogenetically close humans. One vertebrate man, however, says his main concern has become large mammals. Slower on his feet than he

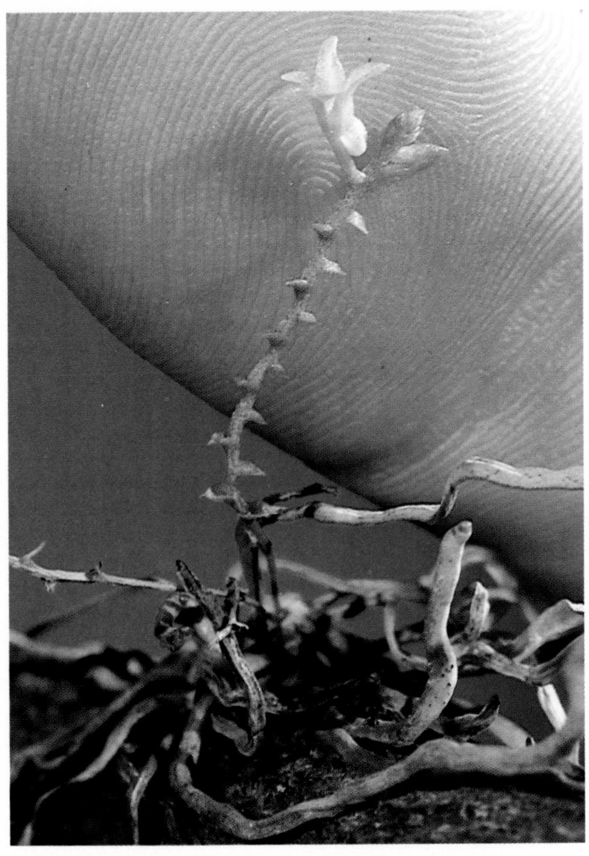

Big, showy and popular prom blooms notwithstanding, most orchids are minute like this one dwarfed by a fingertip—less than half an inch long. This leafless species, an "air plant" or epiphyte, which clings benignly to a host tree, contains vital chlorophyll in its root structures.
Opposite: *The orchids constitute one of the largest, most diversely adapted and intriguing families in the plant world and can, in fact, be found almost anywhere plants grow naturally.*
Pages 144–45: *Orchid fanciers around the world have raised countless hybrid orchids, like this member of the genus* Cattleya.
Pages 146–47: *The orchid family—or families—comprises perhaps 30,000 species found everywhere that plants grow. Many are epiphytes and cling to other plants above the ground, though none parasitizes its host. Still others, like fungi, live on dead organic matter without benefit of chlorophyll. The whole group is considered endangered because many of its hard-to-identify members—like this* Dendrobium *from Southeast Asia—are vanishing in the wild.*

once was, he no longer risks treks across South American pastures where he might encounter an angry bull, but walks around instead.

The closest scrapes Museum scientists talk about have nothing to do with science. Marine invertebrate curator David Pawson lost a colleague with whom he'd discovered and named a new species of sea cucumber. The man was scuba diving in the Caribbean, studying animals that apparently shared a habitat with a cache of drugs; smugglers dropped explosive charges around him while he was in the water. Half the world away, a Museum entomologist had a collecting trip in Sri Lanka interrupted by an insurrection. He checked with the local police who said it would be safe to travel and almost believed them despite sniper fire and road blocks—until he passed a police station in smoldering ruins.

Conservation of scientists aside, the Museum's men and women are dedicated to the conservation of the natural world, and frequently speak eloquently in its behalf. A serious threat to both science and the planet lies in the destruction of mass habitats, especially in underdeveloped nations. The ornithologists and entomologists are particularly concerned: birdmen because migratory species need stable wintering grounds as well as nesting territories; the insect people because so much work remains to be done simply in terms of cataloguing the world's arthropods.

In this respect the Smithsonian plays an international role. Curator George Watson, for example, helped negotiate the Migratory Bird Convention between the United States and Japan, a compact to protect species that visit both countries each year of their lives. The list has awesome dimensions: "three loons, two grebes, eight petrels, five storm petrels . . . boobies, frigatebirds, egrets, a wallow of ducks, geese and swans, three hawks and eagles, the osprey, three falcons, a wandering of shore birds, skuas, avocets, two owls, nightjars, woodpeckers, swallows, finches, the skylark, flycatchers, warblers. . . . " Almost all families but the crows.

Secretary Ripley also wears the hat of President of the International Council for Bird Preservation, the outstanding bird protection organization in many people's opinion. In conjunction with the Smithsonian it publishes the Red Data Book, a multivolume catalogue that covers birds, plants and other organism groups. In addition, several years ago the

"Lady of the Night" (Brassavola nodosa) *thrives in Central America, for the time being at least. Many nocturnal orchids evolved in association with night-flying insects and bats, which feed on their nectar and pollinate the flowers in the process. Highly evolved orchid species display remarkable ingenuity in their reproductive mechanisms but consequently depend on very specific habitats and faunal allies. Thus many species are highly vulnerable to environmental change.*

The family known for exotic elegance also includes such seemingly rustic flowers as the yellow lady slipper of the Appalachian woods.

Museum established an Office of Biological Conservation which is particularly active in worldwide botanical matters.

In terms of the collections and their safekeeping, the Museum presently suffers through a transition period which threatens to continue. After years of planning, the Museum Support Center will open for business in nearby Silver Hill, Maryland. Nobody speaks of "moving day" here because with several dozen million objects to transport—many of them irreplaceable—the process of moving will take at least five years.

Reasons for the move are manifold. Simple overcrowding was one. The two wings built in the 1960s removed some pressure temporarily, but within two decades the Museum's seams were bursting again. Antlers and mounted mammals were being stored in an attic where temperature and humidity graphs must have looked like roller coasters. The Rotunda's upper gallery housed banks of human skeletal material in drawers 14 tiers tall. Lacking a room of its own, the fabled Terry Collection was kept in a hallway with banks of file cabinets arranged as four walls enclosing the cramped workspace. Primitive pottery was stored in another attic, extinct marine mammals in a dank basement. A dim corridor near the Director's office was pressed into service as a storage range for small moths, a place so poorly lit that one technician habitually worked with a miner's helmet and lamp on her head. Two exhibit halls were "borrowed" from the public arena and turned into doubled-decked laboratory, office and collections space for entomologists. It came to pass that there simply wasn't enough room. Overcrowding aside, a turn-of-the-century building is a poor place to store objects that can be harmed by dampness, erratic temperatures and sunlight; things mold, rot and oxidize.

The new Support Center, containing 300,000 square feet of floor space beneath a five-acre roof, has offices, conservation laboratories, study rooms, research facilities and, set apart, huge storage ranges. The separation assures better conservation in many senses. Things will be stored here exclusively; other work and human traffic can be kept to an absolute minimum. Technicians won't bring in their brown-bag lunches or install potted plants to attract (or transport) vermin. Atmospheric conditions will

be finely tuned; the contractors promise a constant 70-degree temperature and 50 percent relative humidity (both figures plus or minus two) throughout the year. While other Smithsonian museums will use the Center, which includes a conservators' training facility, nearly three-fourths of the huge space is dedicated to Natural History. The Museum people hope it will be a showcase facility, the ideal architectural environment for the conservation of all these objects.

The passenger pigeon, "most gregarious bird ever known," became extinct in 1941. But visitors see their lifelike remains in an Appalachian diorama here and wonder at the lost spectacle of a flock of a million migrants darkening the autumn sky.

THE MUSEUM OF SCIENCE

In scientific terms "increase of knowledge" translates into a single word: Research. More of it goes on in the National Museum of Natural History than in any other building on the Mall. Among Smithsonian bureaus, several do nothing but research in far-flung places, while other of its Washington edifices serve primarily as public showcases. This Museum attends to both knowledge's increase and diffusion because despite Joseph Henry's misgivings, the Smithsonian in general and then this building in particular became a "collection of collections." The objects demanded and inspired investigations. The collections also had to be catalogued and curated, organized and preserved. This required the professional attention of special kinds of scientists and research people. Over the decades the business of caring for myriad objects demanded and stimulated basic examination of what exactly they all were. The collections grew in part because interested specialists came here, often bringing their own material with them or bequeathing it later. Whichever was the inquisitive chicken or the acquisitive egg, the Museum became an international center for research in several fields, especially taxonomy, mineralogy, petrology, physical anthropology and North American ethnology.

In some respects as it pursued these ends, the Museum became unique: neither a university nor a privately endowed institute (but somewhat similar to both), it is a stepchild of government devoted to that most impractical pursuit, pure science. The goal here is *not* to feed the hungry, conquer disease, end the threat of war nor reach any other specific, immediate, practical goal. It is to perform pure science.

The distinction between "pure research" and "applied science" lies more in motive than method. (Indeed methods may be identical.) A Museum research scientist seeks new information about some facet of the natural or cultural world and insight into it. By contrast a drug company geneticist or oil company geologist intends to turn data into something concretely useful and commercially profitable. Museum scientists don't deny the possible applicability of their work; they do assiduously squelch thoughts about applications and gainsay any practical motive when deciding what to study.

Each of the nearly 250 senior scientists insists he or she pursues a

A mosaic of scanning electron microscope (SEM) photographs enlarges a sea fan's spicule 7,500 times—almost the size of a table top. The organic tissues of soft corals and sea fans are filled with these minute calcified structures, whose shapes and architecture provide the basis for classifying the families and genera and species. The SEM's fantastic magnifying power permits taxonomists to explore previously inaccessible frontiers, such as the crystal microstructure of this spicule, as they pursue the process that is science in the "Museum of Science."

Opposite: *This headdress and nose ornament are made from the feathers of numerous male birds of several Amazonian species. They were worn on festive occasions by a man of the Waiwai tribe, which lived in the forests of southern Guyana. Like the female birds, Waiwai women possess fewer and less conspicuous kinds of ornaments.*

scientific interest "for its own sake." That shibboleth denotes a necessary safeguard against investigative bias and highlights the intellectual luxury abounding here: the wherewithal to pursue a line of questioning without the distractions and pressures of eventual utility. By tradition and explicit agreement, the value of museum research does not depend on such practicalities as inventions patented, lives saved, dollars earned. The science is its own reward; the pure scientist's goal is to spin a new thread in the web of human understanding. Just how a new discovery or principle may be applied—for good or ill—is somebody else's business. If practical applications were the mandatory result, research here would stop in its tracks. If any Museum investigator ever made a better mousetrap, it was not to lure the world to his door but to catch more mice for taxonomic study.

The specific subjects that Museum scientists explore can certainly appear esoteric to the point of absurdity to an outsider's eyes. But a little digging—at the risk of asking hopelessly pedestrian questions—belies the boast of perfect impracticality. Precisely derived knowledge has a way of being provocative and potentially useful. More often than not the implications of original work are simply intriguing; how they may eventually lead to practical fortune remains to be seen. Witness Secretary Henry's network of volunteer meteorologists—organized for the pure purpose of collecting data nationwide—which became the United States Weather Bureau, now a Commerce Department agency that serves us all. Witness a current ethnological study of Indian headgear which curator and editor William Sturtevant ponders between deadlines for the 20-volume *Handbook of North American Indians.*

The headdress in question is the swept-back eagle feathered type worn by the rebellious brave in every Technicolor western I ever saw as a boy. Art (if those oaters can be called that) was mimicking life here, yet life had mimicked art long before that. Contrary to popular belief and some anthropological consensus, this investigator now suspects that these headdresses were not native to North America, only to the southern continent. Feathered garb, indigenous to pre-Columbian Brazil, reached North America circuitously—via the salons of imperial Europe. Yet the borrowed attire became as typical for northern Indians at ceremonial occasions and

The Hopewell Mound people, who made this disk, influenced much of middle North America from their centers in the Ohio Valley. The stone's purpose has not been discovered, though stains on later ones like it suggest they served as paint palettes. Found in a cemetery—typically the mounds were such—it displays a sophisticated artistry combining realism with abstraction. Excavations of villages show that these hunter-gatherers occupied established settlements for generations, practiced some agriculture and thrived for fully a millennium before disappearing circa 600 A.D.

Opposite: *Frontier artist George Catlin bought this cradle, he claimed, "from a Sioux woman's back, as she was carrying her infant in it." He probably acquired it during one of his summer trips to the Eastern Dakota of the Upper Mississippi Valley in 1835 or 1836. It is an outstanding example of Indian quill work in which both life and geometric forms are used.*

public events as rented soup-and-fish among waspish fathers of middle-class brides. Call the phenomenon a case of sartorial migration or an intercultural pas de deux.

Illustrated accounts of the earliest explorers' discoveries in the Indies won instant popularity among European literati in the sixteenth century. The first published picture of Indians, circa 1505, depicted men wearing crowns of upright feathers (and feathered skirts) in a woodcut whose specifics are confirmed by literary evidence and reality in later centuries. The subject was Brazilian, a detail soon lost in the shuffle of European curiosity about the New World's inhabitants. A decade after the woodcut appeared, Albrecht Dürer published a similar engraving that was widely copied. Within a century the image was fixed in the European mind, which applied it to Indians everywhere and used it to symbolize the Americas. A 1613 tract describing one of Champlain's expeditions contains internal proof of the spreading sham. No doubt Champlain himself saw whatever he saw, including a Loch Ness-like "monster," but the artist who illustrated his journal hadn't. He drew palm trees growing in Mohawk territory, now upstate New York, and Mohawks wearing feather hats and skirts. The preponderance of other evidence reveals that Indians of that region simply didn't wear such garb in the seventeenth century any more than their forests produced coconuts. The artist just depicted Indians as he believed

Opposite: *The Hopi priest in this diorama concludes a nine-day ceremony dancing with a live rattlesnake in his teeth while another priest diverts the serpent's attention with a feather wand and others chant. The Hopis then release the rattler to convey their prayers to the gods, who provide rain in their arid land.*

Left: *The scientific value of any object depends in part on its provenance. By itself this elkhorn implement is a handsome curiosity. But George Catlin acquired it before 1840 from the Sioux and reported that it was a war club embellished with two copper-disk eyes to enhance its resemblance to a water bird. Thus it can be studied within a more precise context to augment knowledge about the people who made it. By a like token, a specimen of an unknown insect species has virtually no value for the taxonomist without information about where it was collected.*

Overleaf: *Within the Arapaho teepee, the host (center) prepares to entertain a guest.*

they should appear; he exercised his own preconceptions and catered to those of his audience. But something else was afoot on these shores.

During King Phillip's War in 1675, the Massachusetts colonial militia struck a brass medal for distribution among Indian allies. It had an appropriate caption regarding the wearer's fealty to the British crown and the image of an Indian with the feather headdress. The Indians who won the amulet wore it with pride. A similar image, slightly modified, adorned a silver pipe Governor William Henry Harrison presented to Ohio tribes in 1814. The headdress began to take on a rakish slant by the early nineteenth century when Charles Bird King painted a Pawnee wearing it; in the century and a quarter since its introduction to New England, it had leapfrogged across the Mississippi. Samuel Morse, wearing his painter's hat, plagiarized King's portrait and placed the man in the House of Representatives gallery for his epic panorama of Congress at work. By the 1830s, Indians of the Great Plains were wearing it more commonly. George Catlin, who accurately painted hundreds of Indians in the West in the 1830s, shows only one of them wearing the feather bonnet which was

Opposite: *Early Americans fashioned stone projectile points in various styles which have been named for the places they were first found. From left to right and top to bottom these are: Alberta, Dalton, Folsom, Clovis, Agate Basin and Clovis again. Anthropologists suggest that these weapon tips were once attached to either lances or spears, but not used as arrowheads. The projectile points pictured here were used to kill and possibly assist in the butchering of now extinct ice age mammals.*

on its way to ubiquitous status. In fact Indians of many nations adopted the headgear, which became the identifying symbol of and for Indians during the 1860s and 1870s throughout the western United States. The device spread back to New England, New York and Virginia, and forward to California and the Southwest by this century, and finally reached the Florida Seminoles in 1950.

In a word, the feather headdress became universal among North American Indians in the course of four centuries. It spread like another exotic, the honey bee which never buzzed these shores before Europeans brought it here but then spread throughout the temperate continent. Sturtevant makes the migration of the feathered hat sound almost inevitable. The European artist who drew the first portrait of a New World aborigine accurately portrayed the Brazilian in feathered dress. Thus a prototype was born and a stereotype inspired. It quickly became the costume that Europeans expected the Indian to wear; a picture wouldn't be "right" otherwise. In time, Indians inevitably saw how the encroaching establishment depicted them; they then emulated the image—just as motorcycle riders began wearing black leather jackets and Marlon Brando hats after *The Wild One* first romanticized the anti-hero. As pan-Indian sentiment grew in the wake of white domination, Indians sought commonalities to distinguish themselves, to declare their common separateness.

When the Museum anthropologist publishes his research, the paper is not likely to contain much discussion of the implications of painstakingly gathered data. Rather it will focus—quite properly according to scientific practice—on the documented specifics of static European perception and changing Indian dress. The study's purpose will be to detail the counterpoint between image and fact as exemplified by feather headdresses. Informally, however, the scientist willingly admits a tangential connection to the broader business of human behavior. Each of us displays specific symbols—in clothing, speech and mannerisms—which contain concrete meanings and carry intrinsic messages. Or so we all believe. But the meanings only seem immutable. Each symbol has its own history which may be as convoluted as the blood feud between the Martins and McCoys or a history of the Balkans.

Using a piece of deer antler, Dr. Dennis Stanford chips flakes from an obsidian blade to replicate a Stone Age Indian tool. He has already used the walnut-sized hammer stones and large sections of moose horn to rough out the "bifaces" before him. What these black obsidian and light chert implements will become—spear-points, arrowheads, knives or scrapers—depends on slight modifications and the fitting of specific kinds of handles. Dr. Stanford experiments with materials like these to discover how archaeological objects could have been made and to understand the potentials of very early technology.

The symbols that societies contrive are typically protean. Initiated by one group under one set of circumstances, a symbol evolves within a changing context—as, no doubt, the Brazilian feather garb had evolved among its wearers to the complexity it had reached by 1505. Transplanted into another cultural environment, it takes on other meanings. In scientific parlance the widespread adoption of the feather bonnet exemplifies "ethnic boundary maintenance." Indians of different North American groups—not all of them amicable to each other until a common adversary appeared—adopted a symbolic article of apparel which distinguished them from their ethnically distinct invaders. Whites became dominant; the borrowed headdress became an identifying mark. Invented by people whom Amerindians had never seen, and introduced by those who became domineering adversaries, it acquired a single meaning in the eyes of the people who wore it and those who held sway over them. It identified the wearer, whatever his tribe, as an "Indian," an ethnic category that didn't exist among this continent's inhabitants before the European invasion. Importantly, it denoted "Indian" to its wearer and everybody else, namely non-Indians.

The headdress became popular among Indians because it came to express a tacit sentiment even before they might have articulated ethnic unity and prideful distinction. Seen in this context, the data present an interesting aspect of pure anthropology. But consider the opportunity it offers a haberdasher catering to a group bent on heralding its own collective identity—teenagers, for example. If a jeans designer adroitly employed the "boundary maintenance" principle (as some seem to) and made a million dollars, Sturtevant's study might be called the keystone of a revolutionary marketing phenomenon. The clothing industry would be incongruously beholden to the Museum's Anthropology Department, as so much of the nation is invisibly indebted to Smithsonian science. However, like most other scientists here, Sturtevant has little interest in modern clothes. He belongs to the fraternity of "pure" and sartorially disinterested scientists.

One of the apparently "practical" programs in the Museum, a metamorphosis of the Southeast Asia Mosquito Project begun in 1964, conceals its theoretical "purity" in an exhibit hall converted to two tiers of labs,

Museum archaeologist Dennis Stanford, an expert on the technology of early Americans, examines a primitive knife (above) *before testing a new hypothesis on an elephant that died in a Boston zoo. Working at the National Zoo's Research and Conservation Center in Virginia, he and his colleagues used spear-throwers* (atlatls) *to show that spears could penetrate the thick hide enough to bring down this kind of game. Then they found they could butcher the animal with bone tools attached to monitors* (below) *which recorded data for wear-efficiency analysis. Nearly 12,000 years ago Indians used stone weapons to hunt mammoths (whose closest surviving kin are elephants). Dr. Stanford has found much earlier kill sites containing not stone implements but bone chips, which he believes may have been used as butchering tools. His field experiment provides analogues to explain how these mammoths might have been slain nearly 40,000 years ago. The history of man on this continent must be revised if he proves that people hunted pachyderms here that early.*

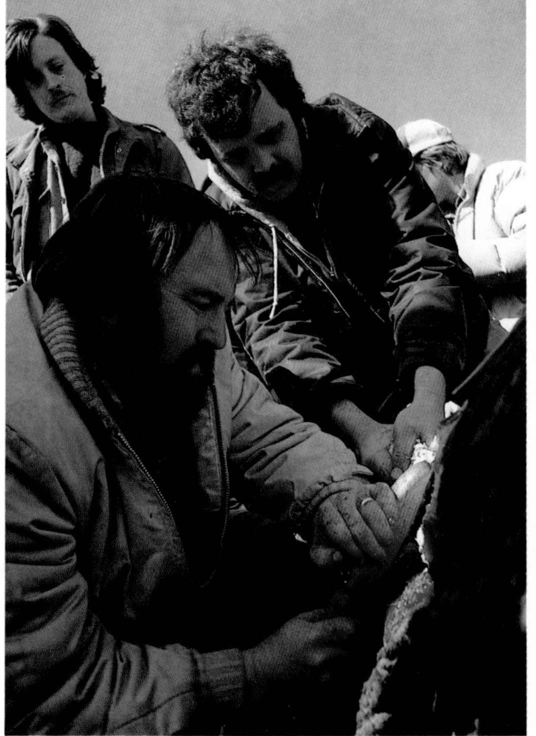

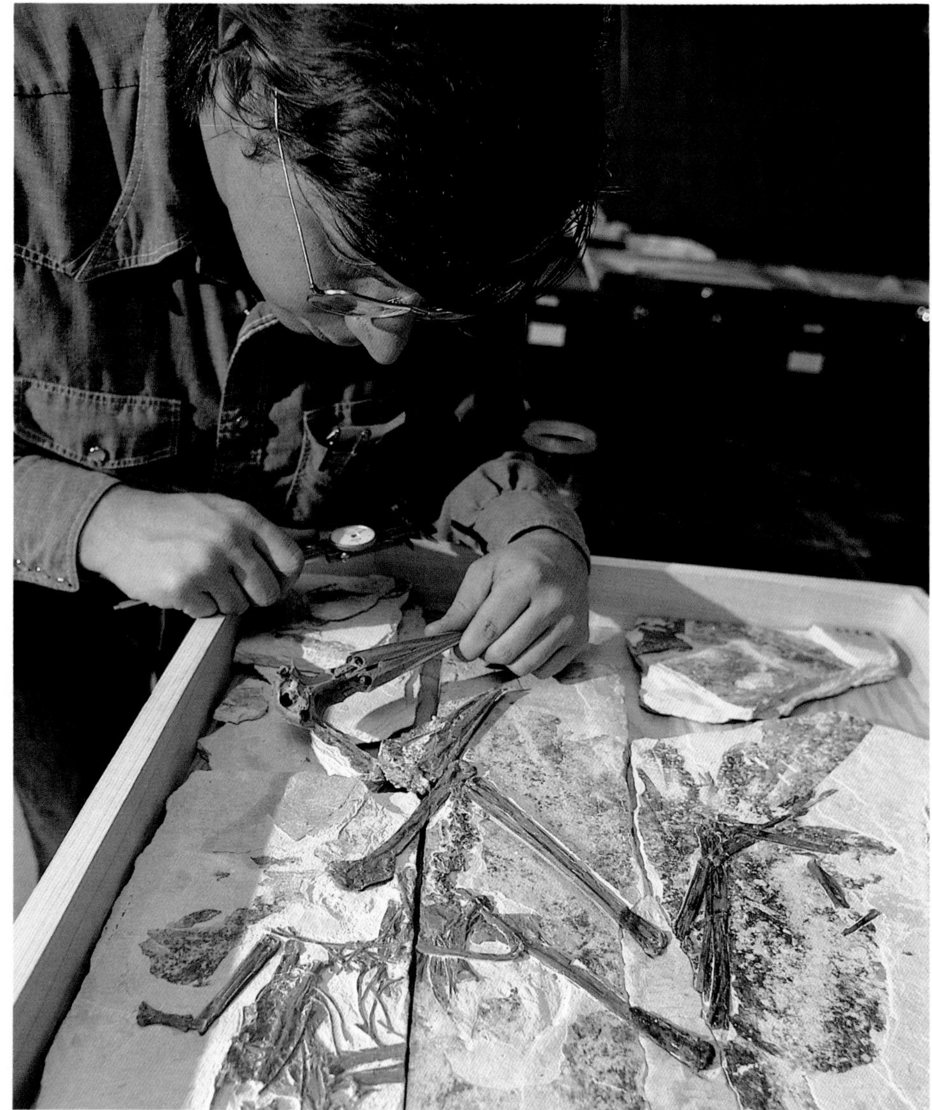

Avian paleontologist Storrs L. Olson studies the remains of the earliest frigatebird, which he discovered and named Limnofregata. *Its fossil skeleton revealed two kinds of relationships: those linking this species with the modern frigatebirds and those linking it with the ancestors of modern boobies, tropicbirds, cormorants, pelicans, et al., all members of the Order Pelicaniformes. This kind of research slowly illuminates the descent and kinship of many organisms in every major taxon and thus outlines the history of life on earth.*

offices and storage ranges. An adjunct of the Entomology Department—the Insect Zoo is just a short crawl away through the Reptile Hall—the Medical Entomology Project is the boarded-out child of the Army Medical Department's research arm at the Walter Reed Hospital complex. The motive behind its "goal-oriented" purpose of identifying all the world's mosquitos is clear enough. Since the dawn of war, almost every army to take the field for more than a week has lost more casualties to malaria and other insect-borne diseases than to enemy forces. The motive is humane as well, since mosquitos show no more respect for uniforms than mufti or naked skin and have decimated civilian populations as well as military ones throughout history.

Whatever the Vietnam Era origins of this unit, the Defense Department wisely doesn't order the MEP to zero in on mosquito species native to potential trouble spots around the globe. The work here involves basic taxonomy, the identification and description of unnamed species, which typically takes longer to conclude than any international contretemps in modern times. Thus three Medical Service Corps officers, indistinguishable from their civilian colleagues, march to the cadence-count of science, systematically studying the family Culicidae. It comprises at least 6,000

Curator Clayton Ray examines the naturally freeze-dried remains of an extinct musk ox. Dissection of the specimen, probably a female, may settle an old controversy about two kinds of musk oxen that inhabited North America during the Pleistocene: were they separate species or male and female of one dimorphic species?

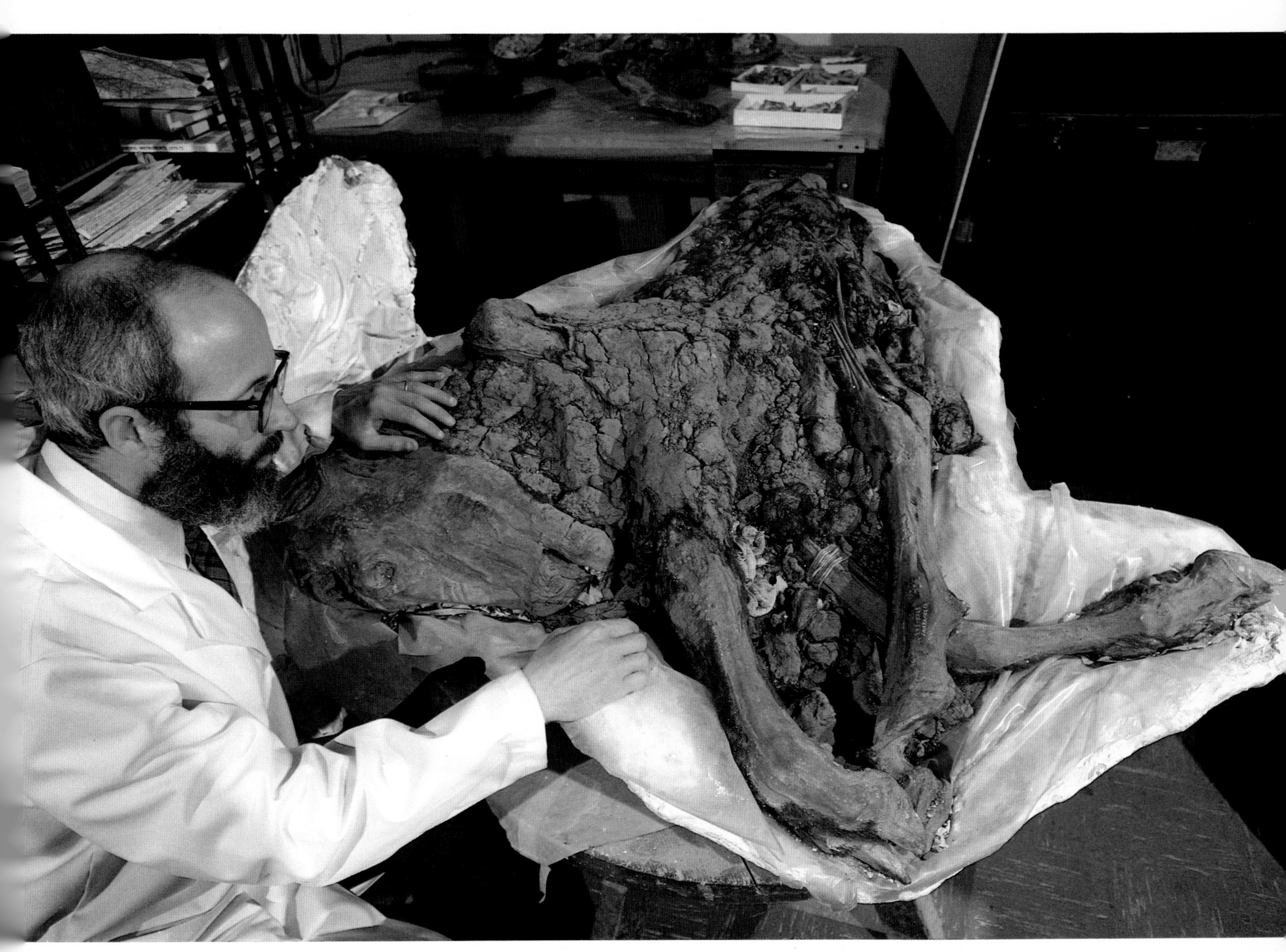

Unique, the black skimmer feeds on the wing both night and day by plowing its lower bill through murky water and snapping up unseen minnows. Now Curator Richard S. Zusi explains one of its anatomical oddities: a split pupil found in cat's eyes but in no other bird. Formed by two muscle crescents which overlap top and bottom, the split pupil opens as wide as a round one for vision in total darkness. But in bright sun, it closes down to 5 percent of its maximum size, four times smaller than the pupils of related birds which fish only by day.

species of which barely half have been named. Of those, more than 200 are known (and several times that suspected) to carry human diseases. It seems a small number of vector organisms until all the protozoans, nematodes and viruses they carry are considered: those causing Rift Valley fever, yellow fever, dengue fever, encephalitis, elephantiasis, malaria in its many forms and more. After 15 years the MEP appeared to have reached the midpoint in its work. But it's safe to assume the goal of identifying all won't be reached before the end of this century. For one reason, new species continue to differentiate themselves from old ones. For another, nobody suggests every culicid habitat has been located so the small army of collectors which gathers specimens around the world cannot be perfectly deployed.

The living mosquitos which those collectors supply keep a staff of technicians busy coddling their charges better than nature does. Females, for example, must be catered a blood meal before they'll lay eggs. The eggs must be carefully tended under controlled conditions until they hatch, then each larva placed in a closed vial to pupate. After each metamorphosis the molted exoskeleton is retrieved, numbered and prepared for microscopic study. When adults appear, each must be isolated and given a raisin or currant for food because male genitalia—the main diagnostic character in many insect species—cannot be conclusively recognized for two days. During that time the terminal body segments change and rotate 180 degrees before the telltale parts become finalized (as it were). Only then can the specimens be preserved for productive study. But the killing must occur on time because some adult mosquitos are very short-lived. Consequently when insects are being reared someone's on duty every day.

The mosquitos perish that the Army entomologists might publish authoritatively on their discovered taxonomy and bionomics, namely their ecology and behavior. Nobody from the Surgeon General on down expects Captain Michael Faran's recent monograph on "The Albimanus Section of *Anopheles (Nyssorhyncus)*" to be a best seller. In fact very few people will ever read it, and few of those cover-to-cover. It is hoped that it contains particular new knowledge for diffusion among men, whether military or civilian entomologists.

The question remains, why doesn't the MEP examine only those mosquitos that carry human diseases? The reasons intermingle the properly ponderous process of authoritative research and the complexity of these sanguinivorous insects. It often appears that a seemingly innocent insect has no medical significance until a human scourge appears or a newly discovered species has been closely studied, then compared with and distinguished from its relatives. This inevitably involves determining apparent relationships among species and other groups in the taxonomic hierarchy (of which more in a later chapter). It goes without saying that closely related organisms often have more in common than distant relations. (Thus a spin-off species of an old flu virus causes an epidemiologically different new disease, i.e., Asian flu, Russian flu, swine flu, etc. Thus the same organism causes German measles in *Homo sapiens* and distemper in other mammals, especially the canids and cats. Thus two members of a single taxonomic class can be infested by the same arthropod to their mutual disadvantage, as in the case of the plague-infected flea which will inhabit a human head if he can't find a warm rat.)

Axiomatically, when studying a group of organisms the group must be defined by organic criteria, not extraneous ones, and for comprehensive results all members of the natural group must be studied. To try picking and choosing only "medically significant" mosquitos would be a hopeless chore with only temporarily valid results. One might just as well draw one's family tree and include only the "interesting" ancestors, say Democrats, bachelors and horsethieves, then let posterity fill in the missing names. Gibbon would have been as well advised to consider only events which occurred in years with a V in them when writing *The Decline and Fall of the Roman Empire*. Thus the MEP has a practical purpose but must observe the impractical rules of taxonomy in order to reach its goal.

Moving deeper into the terra incognita of "pure" research, consider studies currently examining a structure found in one group of worms, animals that possess an unexpectedly enormous number of distinctions from each other and the rest of us.

All that wiggles aren't worms and all worms don't wiggle alike. Common earthworms and nightcrawlers, for illustration's sake, can move

Left: *Can evolution work "backwards"? The golden marmoset lacks a feature famously possessed by all other primates: the opposable thumb. Its loss of the crucial hand muscle represents a departure from the group's common ancestry. Further, this rare denizen of Brazil's coastal forests has nominally "primitive" claws instead of "advanced" nails. Curator Richard Thorington suspects these changes occurred as the animal adapted squirrel-like ways after leading the more monkey-like life of its relatives for millennia. This example of "regression" challenges the conventional tenet that organisms become more highly specialized as they change.*

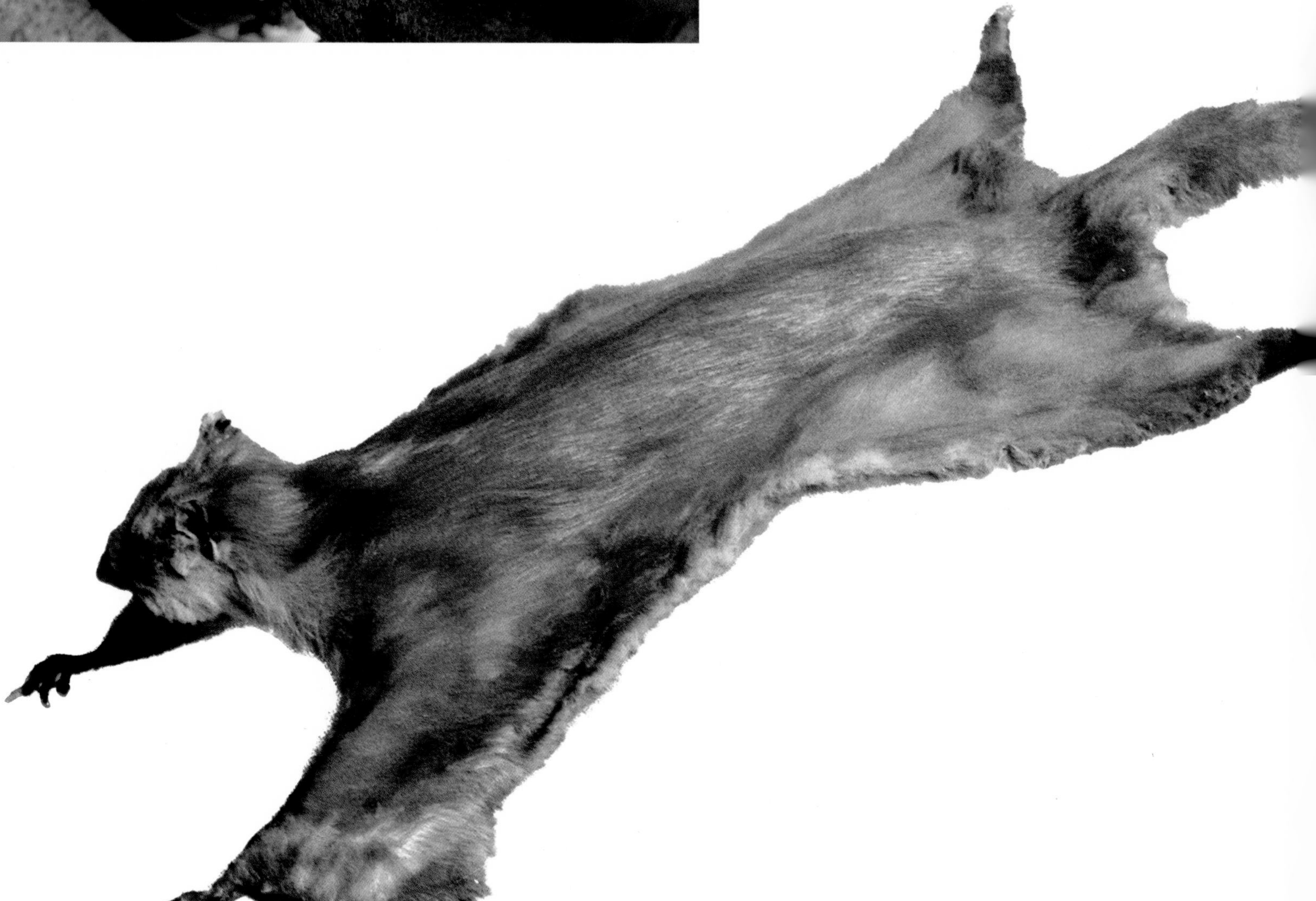

from side to side and undulate vertically or diagonally as they negotiate an unkempt putting green. Far less versatile, members of the Enoplida order of nematodes move about by undulating only along one plane in a sequential ripple motion that resembles a wave moving across the sea. (Parenthetically these little animals, which live in wet soils and saturated sediments, do something closer to swimming than crawling. Almost radially symmetrical, they're as comfortable upside down or sideways as rightside up. The physiological fact remains: They can only wiggle through the ventral-dorsal plane, not through the anatomically lateral plane.) The reason for this limitation lies in their anatomy. Two bands of muscle stretch fore and aft along their entire length, one along the dorsal side (the "back" if they oriented themselves like snakes), the other along the ventral side (analogous to the belly of a reptile). Lacking the muscle bands that girdle earthworms, they have no muscles along their "sides" so lateral motion is impossible. How they coordinated their rippling-wave motion had been a mystery until a Museum scientist addressed the question.

Two decades ago, W. Duane Hope explored the lateral fields, the two nonmuscular regions that separate the Enoplida's ventral and dorsal muscle bands. One function of this tissue is to secrete material for the cuticle which covers the body with a flexible tube that is molted periodically to allow growth. In addition, the lateral fields contain what look like long, serially arranged filaments whose existence and purpose had been overlooked. Hope wanted to study these structures and their odd swellings in minute detail, but another scientist examined them under a light microscope and published some initial findings which obviated immediate further study. Hope turned to other parts of other nematodes and kept productively busy for years, pursuing the arcane science of vermiform taxonomy.

The laws of aerodynamics apply to mammals and monoplanes alike. Petaurista petaurista *has a membrane between its hind legs and tail which its smaller kin lack. Heavier, it must attain higher speed—about 25 mph—to glide. But landing at that speed would break bones, hence the extra membrane to provide extra lift, enabling the animal to "stall" at a slower speed and lose momentum before touchdown. Early airplane designers discovered the principle when building training craft: the wider the wing, the lower the flying speed. A taxonomist studying these Asian squirrels used aircraft and hang-glider theory to explain morphological differences between species.*

SMITHSONIAN INSTITUTION
ASOR
BAB EDH-DHRA, JORDAN
CEMETERY A
TOMB NO. 100 N
13 JUNE 1977
A100 N 1
A100N 2

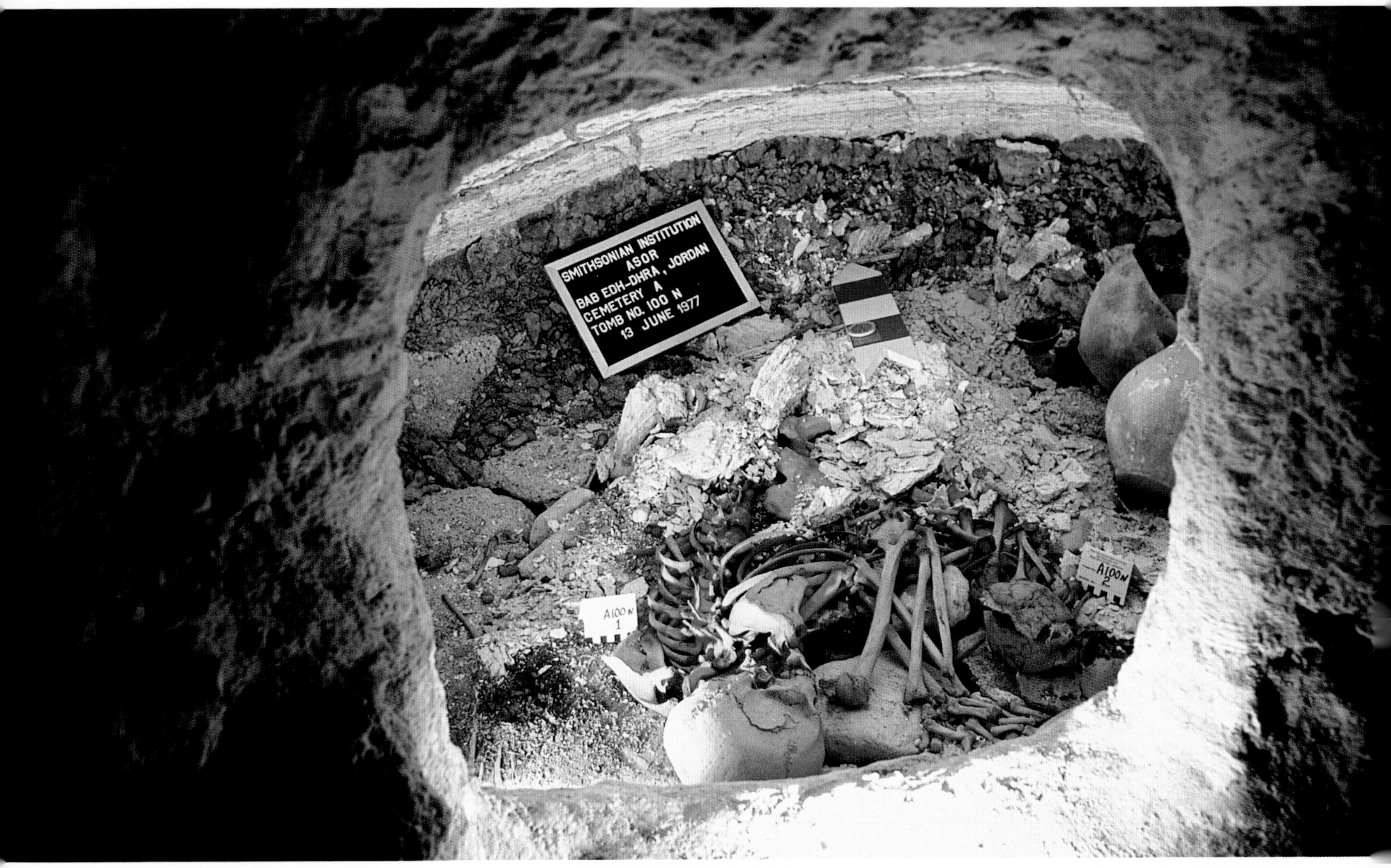

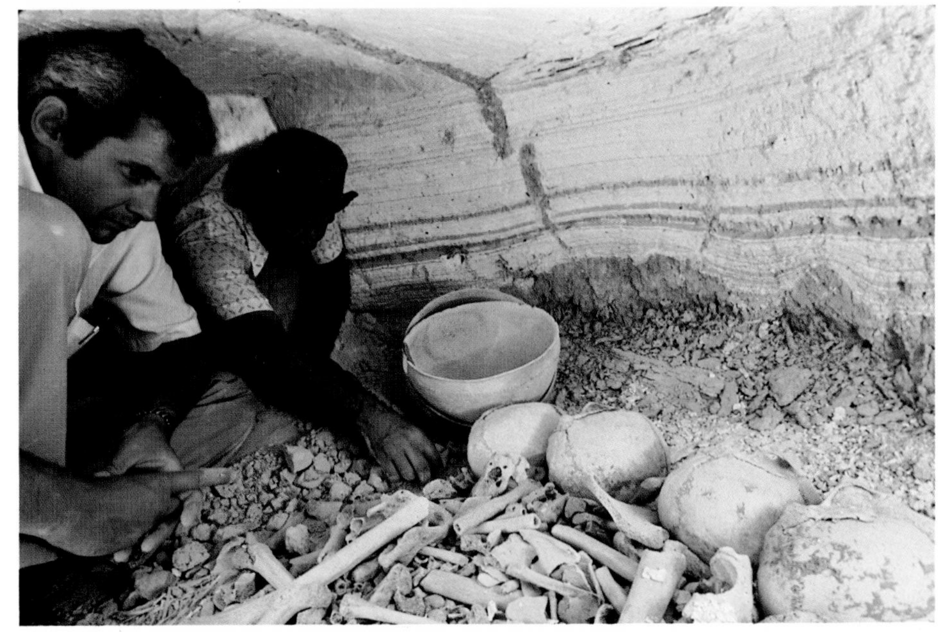

When he returned to the mystery of the Enoplida mobility, he took along with him an electron microscopist, his research associate pro tem at the Museum. Hope and Dr. Stephen Gardiner found much more in the lateral fields than had met the eye of less fastidious investigators. These worms measure less than three-quarters of an inch in length. Using marvelously precise equipment, the investigators sliced them into sections barely 1/250,000 of an inch thick (0.1 microns). Using the electron microscope's enormous magnifying power, they focused on the long filaments punctuated with spindle-shaped swellings, which had been taken for nerves. Not at all, Hope and Gardiner discovered. At least they weren't sensory receptors of any ordinary sort previously seen by science. Rather, these structures comprise paired cells that fit together like costume jewelry poppit beads or stubby jacks and sockets for phonograph speakers. In each pair, one cell has a hairline cilium which plugs into the matching recess of the other cell. Arranged almost in rows, these pairs of cells stretch the length of the worm's body.

Preliminary work suggests these pairs of interlocking cells may serve the same kind of function as ordinary sensory receptors in other animals though their architecture and hence their mechanics are very different. As this worm moves by undulating up and down—as the ripple-wave passes along its body—each cilium in sequence must disturb its receptical cell. Thus the paired cells monitor the passage of the rippling wave as it passes along the body. (The receptical cells probably transmit information about the degree of body bending to the central nervous system in reasonably conventional ways.)

If this preliminary thesis proves out, the work will have revealed a previously unknown kind of sensory receptor, one not known to occur in any other animal. In common parlance, this represents a discovery, though scientists always semantically hedge their bets. Are these cell pairs "unique"? Who can say anything more than that they've never been seen before. "Previously undiscovered" fills the bill and who knows where they'll show up next, hypermicroscopically speaking? It often happens that once one pioneer describes a new kind of landmark on the frontier of his specialty, other scientists recognize it in their own realms. This is how science progresses.

Preceding pages: *The Museum's re-creation of the Bab edh-Dhra burial chamber.*

Opposite, top: *An actual burial chamber in Bab edh-Dhra near the Dead Sea reveals human remains and worldly goods which lay here undisturbed from about 3050 B.C. to A.D. 1977. When opened, this crypt contained food jars, stone scrapers and wooden implements. Some chambers may contain weaver's tools. These burial chambers provide clues to the lives and culture of their owners. Some of the bones show that these Early Bronze Age people suffered such maladies as tuberculosis.*

Opposite, bottom: *Curator Donald J. Ortner and Jordanian colleague Sami Rabidi inside a chamber dated 100 years older (3150 B.C.), in which the arrangement of bones proves final interment in the chamber subsequent to burial a few years earlier. The people of Bab edh-Dhra, who buried their dead in these chambers, may have been more nomadic than their sedentary successors who lived in the archaic city and buried their dead nearby immediately after death. The cemetery included an area of about one square mile and probably contains more than 250,000 skeletons. The chamber seen in this figure is entered from a central shaft which provides entryway for as many as five chambers. In later times people were buried in mud-brick burial houses.*

Opposite: *Animals both fanciful and real —a seal (upper right) and a walrus (lower right)—adorn this walrus tusk which a Bering Sea Eskimo carved early in this century. Eskimos have been carving ivory images of strange mythological beasts for more than 2000 years. Originally they used these beasts as hunting magic on weapons, but by the twentieth century they had adapted their work for the growing European-American market in Eskimo art.*

Hope and Gardiner are certain enough of their findings to start formulating a new hypothesis about the functional anatomy of these worms. In its testing-out—at the Museum and the handful of other laboratories that study Enoplida—other questions will arise: How does this mechanism affect the entire animal? What other animals possess similar equipment? If some do, how might they be phylogenetically related to the nematodes? How did this odd sensory structure evolve—from external cilia in an ancestor perhaps? (In many small organisms, larval mollusks and human sperm cells for example, cilia serve as the only means of locomotion.) Does it represent an evolutionary advance, a primitive anatomical condition, a developmental dead end, a missing link? These questions will not be answered immediately; indeed all the pertinent questions about Enoplida's newfound neurology cannot even be intelligently formulated yet, and therein lies one of the distinguishing elements of pure research. It rarely leads to fortune, only to new questions.

In applied science, of course, the importance of a new discovery or invention is immediately clear. When three gifted tinkerers at Bell Laboratories contrived the first primitive transistor, they could reasonably predict that it would soon revolutionize practical electronics. The "pure" scientist's discovery may have no application or implication in his lifetime; the discovery in and of itself must be its own reward. Indeed any specific advance in theoretical science may never have any palpable significance outside itself.

This truism leads to an obvious question. Why does the Museum pursue pure science? Or, to get even more basic, for what most fundamental reason does this Museum exist? Consider an answer of misleading simplicity: It is a monument. (Its size and shape, its impressive physical existence, declare that it represents something of singular importance to the culture which raised it with pride.) It is also a kind of memorial, although that may not seem so clear.

Like every monument, it is a conspicuous luxury. Its creation and maintenance bespeak the dedication of physical, economic and human resources that could have been put to other purposes or squandered to no purpose at all. That it was built and remains dedicated to natural history

makes manifest the nation's belief that its varied contents—animal, vegetable, mineral and manmade—possess intrinsic or sometimes inexplicable worth. Quite simply, it is a monument to the natural world and a tacit acknowledgment, a statement in granite and glass, of our perpetual wonder and active curiosity.

As a memorial, less obviously, I believe the Museum celebrates man's ancient emancipation from the tyranny of pure biological practicality. Mankind remains biological, of course; we live, reproduce and die. But among the world's fauna we are unique in the enormous portions of energy which we devote to things other than the necessities of food, shelter and reproduction. We can even claim legitimate progress in this: Since the technological beginning, man has been progressively less enslaved by the zoological prerequisites for life. First the invention of the pointed stick, then the plow, freed people from the oppressively constant search for vegetable food wherever it happened to grow. The contrivance of the thorn fence to enclose tractable animals freed man from the uncertain hunt for perishable meat. The mastery of fire and the ability to sustain it at a

Above: *An Indian drum from the Northwest Coast, with a bird in the complex local art style, which often incorporated human attributes in depictions of animals.*

Opposite: *The Alaskan Eskimo's hunting visor had the practical purpose of shielding a kayak hunter's eyes from the sun's glare. It also had ritual significance in its adornments: cormorant and old squaw feathers, carved ivory gull and walrus heads, and wing-shaped ornaments. These devices evoked the spirits of those animals, the walrus and duck which were good for food, the gull and cormorant which were an able scavenger and fisher, respectively. Aided by such allies, the hunter believed his own skills would be sharpened and the spirits of his intended quarry more willing to be captured.*

California's Pomo Indians decorated their gift baskets with feathers. The Kachina doll (opposite), *probably made for a Hopi child, boasts a feather headdress like those worn when Hopis danced in the array of their gods' messengers.*

campsite or cave mouth released our ancestors from unblinking vigilance against prowling predators through the night. The oil lamp, the candle, the electric light ended enforced idleness which dominated the dark hours that are half our lives. Social organization (though not a human monopoly) enabled individuals to concentrate on one kind of work, rely on neighbors for other necessities and engage in the cooperative interdependence that through the growth of cultures became civilizations. Since our species' birth people have gained discretionary time (as economists speak of "discretionary income") to use as they saw fit. Our antecedents won the opportunity to do in part what they chose, not only what they must (such as forage for roots and berries and keep the bears at bay). Throughout human history people have built upon this legacy from the time the original man won the first luxury by putting aside a carcass or coconut to save the trouble of foraging tomorrow.

A wonderful constant of human nature is that mankind has often used the hard-won time so well. *Homo sapiens* is an energetic species, an animal that communicates, cooperates, explores, embellishes and enjoys life in its course. (We are not alone in any of these. Bees tell each other where to find food. Pairs of mating birds share the work of nestbuilding. Wolves in a pack talk to each other during the chase. Otters and dolphins throw themselves into pure play with strenuous enjoyment. Most mammalian females protect their young and males defend their mates.) But man alone, doing all these things and more, found reasons beyond biological ones for expanding the cooperative on impractical activities.

As people prehistorically gained time in their lives, they filled it. Whether Bronze Age Semites or Alaskan Eskimos, all exhibit a common

trait: some degree of dissatisfaction with pure utility and simple sufficiency. A bone case holds ivory needles no less surely if its surface remains bare, yet the people Edward Nelson met on the Bering coast decorated each etui with carvings and visual puns. A plain basket or clay pot holds grain as well as one with decorations, yet the people of Melanesia and Arizona alike took pains to beautify their utensils. (The original motive was proto-religious, no doubt, but esthetics somehow followed.)

Winning time, exploring the earth, embellishing, man has never been content to accept things only as they first appeared. For good and ill, he has been a creative and a controlling animal. We have even made alien forces more familiar by ascribing to sun, thunder or fate the traits of superhuman gods who could be appealed to if not always appeased. Herein the Hopi shaman who carved Kachina dolls out of wood shared an ambition with the dilettante Smithson: Each sought to clarify dim forces according to his culture's means of explaining the world. Each explored the fringe of the unknown and thus nurtured his people's understanding of what they experienced.

What has all this to do with the Museum? Everything. It was not enough for the proto-Hopi to realize that forces beyond human comprehension created and control the universe; he must name those forces gods, endow them with myriad anthropomorphic powers, then carve wooden effigies of them for his children's education. With considerably greater hubris the people called scientists—from before Smithson's day until after our own—examine and describe these forces and other physical entities through the imperfect but testable tools of science. As the Zuni artisan embellished his clay pot according to the ethos of his tribe, so a roster of investigators from astronomers to zoophytologists expand the realms of our understanding and deepen our ken. (Not all Museum scientists, by the way, believe in Science alone. Many are practicing theists, which is to say they belong to formal established faiths. While they render unto science what belongs to the physical world, they render unto religion the inexplicable mysteries.)

In one respect the Museum's scientists are explorers; in another they are embellishers no less than any totem carver. It is not enough to have

Opposite and above: *New World archaeologists have yet to explain the symbolism of this 2,000-year-old hollow pottery figurine from Vera Cruz, Mexico, or the prehistoric Peruvian jug, called a stirrup spout vessel, made by the Nazca culture.*

Overleaf: *The Bantu-speaking Luvale people of west-central Africa hold elaborate celebrations when their boys reach manhood through the circumcision rite. From left, the ritually garbed men are:* Utenu, *a spirit who keeps women away from initiates;* Chizaluke, *a fierce spirit chief; two initiates; the beautifully seductive* Mwana Pwebo; *and* Chikuza, *a good spirit chief. The diorama, featuring authentic costumes and accessories, is based upon various anthropological field studies.*

Above: *The carved cat, a rare find because wooden objects disintegrate, was made for some ceremonial use by the little-known people who inhabited Key Marco, Florida, six or eight centuries ago. While it might appear to resemble Meso-American or Egyptian figures, archaeologists discount any connection between these cultures.*

Opposite: *Made of painted pottery in the Coclé style of Pre-Columbian Panama, a jaguar or "cat demon" bares its claws.*

known for more than a century that life *evolved*; they must learn how it evolved and explain the process step by step. Why? Because curiosity and explanation, exploration and embellishment are among the characteristics that distinguish us from beasts. Because part and parcel of human nature is the habit of creatively filling the time our ancestors won from the purely biological functions that monopolized their ancestors' lives.

Leap if you will from the first aboriginal cairn and its builders to this modern monument, its stewards and denizens. Jump from the student of early science to the Institution he endowed which now conserves collections of Zuni pots, Hopi dolls, native minerals, nematodes and microlepidoptera. Besides collecting specimens and artifacts, the Museum offers men and women a place to investigate the world as we have come to understand it. The work of cataloguing a mountain's fossils may have no more substantive effect than the incised key around the lip of a terra-cotta jar. The phylogeny of one medically harmless mosquito living in some rainforest may in and of itself never generate a useful application or implication. The fact remains that some modern heirs of the proto-people

This Haida pole, with a door opening at the base, depicts "spirit watchmen" and a killer whale, as the armorial bearings of the owner of the house before which it stood.

who first made plows feel moved to practice taxonomy, and others to pursue anthropological riddles. (Our society can support their explorations. Not every man must grow grain. Others build the farmer's tractors. Others still throw pots on electric wheels or spend their days watching soap operas and no one questions why.) Some fortunate inheritors of freedom from pre-cultural subsistence use the time won by our cunning ancestors to climb Mount Everest "because it's there." Well, the echinoderms are there too, and the meteorites, the angiosperm fossils and the relics of stone age folks who didn't differ much from us.

All manner of things in the natural world "are there" to be "conquered" in the mountaineer's vocabulary. More accurately, I think they are there to be possessed or embraced or understood by inquiring, exploring man. Because it is our nature to inquire, examine, explain and embellish. The Apache youth climbed a Rocky Mountain to return as a man with an eagle's talons in his fist. The New Zealand adventurer scaled a Himalayan peak to plant a flag briefly symbolizing his achievement. Both the Bronze Age artisan and the Zuni potter enhanced their people's lives in one fashion. The pure scientist illuminates one cranny of our world in another. It occurs to me that the Museum exists because our people recognize what amounts to a common human trait: the compulsion to enhance our lives, to adorn the plain, to examine the complex, to illuminate the unknown, to penetrate the frontier whether physical or intellectual. The Museum is a kind of temple of the natural world, a monument to natural science. It is a memorial to man's hard-won freedom to naturally explore them both, a tangible testimony of our intention to continue—through scientific research—because that is a part of our nature.

Ten feet tall and wearing a headdress of volcanic rock, this Easter Island stone figure commemorated an illustrious ancestor who brought benefits to his living descendents. Placed on a large stone structure, these sacred places were used for burials and religious ceremonies.

THE PUBLIC MUSEUM

"Diffusion of knowledge." The fusty phrase has an almost chemical ring to it, or an alchemic one that conceals all the fun involved. It stuffily evokes the dispersal of learned discourse through published monographs, the devolution of the latest scientific word through dour gentlemen's colloquy, the summoning of cognoscenti to esoteric symposia, the exposure of doctoral candidates to pure pedagogy. In fact the Museum nurtures all these activities to diffuse knowledge. As a recent annual report revealed, during one typical year Museum personnel published the following: a dozen books (including a definitive 3-volume, 2,735-page inventory of the continent's membrane-winged insects); 32 new monographs in the *Smithsonian Contributions* series, which now runs to nearly 500 volumes; yet another volume in the *Handbook of North American Indians*; upwards of 600 papers in scientific journals. Moving on from the blizzard of published words, there's no way to adequately quantify the Museum's role in stimulating worldwide scientific dialogue, nor to enumerate the scientists who meet with colleagues here, attend small seminars or landmark international symposia. Just as it pursues the "increase" of knowledge through original research, the Museum disperses the fruits of these labors to the scientific world. But if it only diffused knowledge in parochial ways—providing "service to the trade only" as antique dealers phrase it in New York—Smithson's mandate would be only partly fulfilled. Instead this institution caters to the general public more devotedly (and benignly) than a mission church in the founder's day reached out to the heathen.

Clearly it is a public institution in that government revenues and tax-deductible donations support its operations. Of at least equal significance, it is public in that it actively serves the scientific laity—ordinary people. The heart of the building pulses with our voices and footsteps 364 days a year. The huge exhibit halls opening off the ceremonial foyer were designed to accommodate—indeed to impress and edify and boggle—everyone who enters. While a very few insiders look up from their microscopes and down their noses at what they call "public entertainment," the vast majority of curators and technicians willingly recognize the Museum's responsibility to serve the public—to diffuse knowledge

Seen from the foot of the dome 125 feet above the Rotunda's floor, a few of the 5 million annual visitors enter the Museum.

Overleaf: *A popular and frequently photographed attraction, the tiger which once stood near the north entrance now leaps in perpetuity at invisible prey near the Museum Shop off the Rotunda. This individual was shot in India after it began slaughtering cattle and attacking people. Later it was given to the Museum of Natural History.*

Opposite: *Below the balcony, the entrance to Evolution Hall reveals a diversity of evolutionary material, from invertebrates in jars of preservatives to portraits of Museum personnel on the tower.*

Following pages: *Visitors encounter evidence of a thousand animals' adaptive diversity in the balcony of Evolution Hall, which features the camouflaged pelts of closely related mammals, the protective colors of various mollusks, the awesome size of a blue whale's suspended skull. The miracle in this sample lies in the apparent fact that all mundane life descended from a common ancestor whose identity defies human discovery.*

broadly among people without regard to scientific background, intellectual origin or academic status. Many of the scientists do it daily as they freely offer advice to everyone from schoolchildren seeking grist for science reports to farmers fighting a new pest to bemused heirs hoping to name the pressed herbs in late Great Aunt Julia's treasured herbarium. Other scientists devote the greater part of their professional time for some years to the planning, design, and installation of an exhibit which may well last longer than they will. As one department chairman puts it, "Any scientist who isn't interested in public education shouldn't be here."

In some instances, the Museum's dedication to the public appears openly seductive, as in the presentation of the lively lectures and free films offered every Friday noon in Baird Auditorium. Other enticements are more subtle. When the Hope Diamond was acquired, suggestions were vetoed to display it incongruously with the elephant in the Rotunda. Instead the jewel was placed in the far reaches of the third-floor Gem Hall. People would seek it out wherever it appeared. Why not be certain that they saw something less celebrated but perhaps more intrinsically interesting en route and in the process learn a new notion or have their curiosity awakened.

If some of the scientists would settle for just exposing people to spectacular objects and hope for the best, others believe they must aggressively teach every visitor something. Endless variations on two themes echo back and forth as each ad hoc committee debates the content of the new exhibit. On one hand: "People want to see *things*—a whole wall of snake skins, an example of every decapod in the crustacean world, every coelacanth in the collection. They want objects, the real things." Colleagues disagree until they're hoarse: "The Museum's purpose must be to educate, to enlighten, not just cater to idle curiosity. We must explain plate tectonics so that they leave the building understanding the mechanics of continental drift." Inevitably neither side holds absolute sway, as a tour of the public halls suggests.

The first-floor space surrounding the Baird Auditorium contains mahogany cases with a mounted example of every bird commonly seen in the Washington region. No didactic argument appears in the labels

GENETIC

aardwolf, red fox (above), jackal, giant panda (skull only), long-tailed coati, raccoon, sun bear, European badger and Alaska brown bear.

BEARS & RACCOONS
(Families Ursidae and Procyonidae)
Bears and raccoons differ from other carnivores in being plantigrade (walking on the full soles of the feet). The broad, flat-crowned cheek teeth reflect the ability to eat many kinds of food. Most species are heavy-bodied and slow in their movements.
The giant panda and the coati-mundi belong to the raccoon family.
EUROPEAN BADGER
RED FOX
Vulpes vulpes
GIANT PANDA
DOGS
mily Canidae)
their relatives are pri-
for hunting over a wide
bringing down prey in a
The narrow, project-
nd the sharp fangs are
prey, in fighting, and in
long legs and com-
haracteristic of running
SUN BEAR
Helarctos malayanus
JACKAL
Canis aureus

Above: *An exhibit may be as close as many people come to an exotic like Indonesia's rhinoceros hornbill. When a female occupies a cavity high in a rainforest tree, she and her mate seal it—except for a small hole—with mud and dung to keep out predators. While captive, she broods and tends the chicks until they're ready to fly. During these weeks the male brings her food.*

Below: *"Birds of the World," a hall that always draws flocks of visitors, features an Antarctic diorama with specimens collected by one of Admiral Byrd's parties. The largest of penguins, the Emperor, here, incubates its eggs and hatches its young in the total darkness and incredible cold of the Antarctic winter.*

Opposite: *A miracle of exhibitionism: feathers in a peacock's tail.*

identifying these relic specimens of the 1920s that were once put away, then re-exhibited in response to public requests. The curved room contains an entire aviary of stuffed specimens and brief, purely descriptive captions; nothing more. Still, people seek out the place—birders sharpening their eye for field marks perhaps and certainly some folks who never ogled a bird in the wild. The geology halls offer a short course in rock science through displays and explanations that are as bare-boned as a fossil shad. Still, people always browse here too. They contemplate the rocks, touch the largest copper nugget ever found, pound the stone base supporting the seismograph machine to see how earthquakes are measured. They leave with a new sense of the world's rocky diversity.

At the other end of the didactic spectrum, the recently installed Evolution Hall explains the theory and mechanics of this most fundamental biological phenomenon, the driving force of the planet's organic diversity. People amble through its compartments to read the explanations of genes and chromosomes, wonder at the depictions of population dynamics among cockroaches, salmon and people. They admire the

Left: *Seemingly bathed in a shower of sparks,* Diplodocus *awaits his domicile's renovation—now completed—as welders construct steel-beam framing for a new balcony.*

radiation of the Ursidae in the brown bear and polar bear, two distinctly named species that can interbreed in captivity but whose habits and habitats separate them in the wild. They gawk at the towering photomontage depicting human variety, a collection of human faces (all of them, incidentally, portraits of Museum people).

Some visitors have objected that the Hall makes no mention of the Bible's account of Creation, although the introductory sign announces the attraction has a scientific message, not a theological or anthropological one. "This exhibit explains how most scientists think the evolution process works." What is evolution? "The concept that species change through time. Over millions of years small changes accumulate to become large differences; new species arise, others die out" like dinosaurs and man's prehuman ancestors.

So the Museum offers raw data in some exhibits and illustrates sound scientific consensus in others. It caters to the visitor who has the itch of specific curiosity and to the person who enters the place for no better or worse reason than Hillary had for climbing Everest: "Because it's there." Specialists, students, sightseers, passersby—all members of the public—visit this public institution and absorb some knowledge from its accessible facets: the exhibits that represent the myriad collections and explain the diverse directions of natural science. These exhibits are, after all, the essential elements that constitute the Public Museum.

These displays result from the co-operation of vulcanologists and painters, cinematographers and carpenters, paleobotanists and plasterers. How they come into being is an absorbing and complex process, even an organic one, to which the Museum dedicates enormous energy. It begins with the continual attention of the standing Exhibits Committee, whose members represent every disciplinary branch of the Museum, plus the Education and Exhibits offices. Looking down the road as far as they can see, this committee roughly schedules the "opening"—actually the complete renovation of an existing space—of one major hall every year or two. Thus the entire public area within the central building will be redesigned on the order of every thirty years, though the life expectancy of any given exhibit is about two decades. Some halls have longer relevance; others lose

Science is dynamic, and Museum exhibits reflect changing ideas. Here workmen dismantle an old display in the Latin American Archaeology Hall (top) *to make room for a replica of an outdoor market in the Andean highlands* (bottom).

Overleaf: *A view of the South American Hall which includes the mural of Indian hunters using the bola. (The mural can be seen in its entirety on page 477.)*

The Mountain Valleys:
Colonial and Modern Eras
Modern Andean Indian culture is a mixture of Prehistoric, Colonial, and 20th-Century practices. This blend of cultural influences is apparent in the scene shown here, the main plaza of a mountain town on market day.

Pages 206–7: *In the South American hall, the sharp contrasts in the environments and raw materials of the Andean highlands (at left) and the Amazonian lowlands (at right) are seen in the dress, foods, ceramics and other aspects of the aboriginal way of life.*

This diorama, depicting a Chinese opera about the salvation of the Ming Dynasty from an evil regent's plot to unseat the infant heir, surprises visitors with a musical recording.

their appeal or effectiveness far earlier for a variety of unpredictable reasons. (In the heyday of slide-show visuals, for example, an ecology exhibit proved too popular for its own good. Depending on a many-projector montage for its impact, it closed early because visitors kept kicking up dust, which jammed the projectors and scratched the slides so often that much of the exhibit was usually "down." Technicians might have tended the too-delicate equipment daily, but only at the expense of neglecting other exhibits in the making.)

Three to five years before a given hall's renaissance, an ad hoc committee, a working group of half a dozen specialists, begins to meet. Comprising curators from a range of relevant disciplines, the task force holds weekly conversations to define an overall concept for the new hall. The first goal is to articulate a consensus about the mega-display's organizing principle in a Theme Statement. One now becoming a reality began: "The basic objective of the Paleontology Hall will be to exhibit the fossil record of life on earth. The emphasis throughout this hall will be on fossils and on interpretation of the fossil and geological record in light of present-day knowledge." That last phrase reflects an inevitable problem: the Museum cannot hope to keep up with the scientific times like a daily newspaper. Any statement silk-screened on a wall has a certain permanence to it. Yet scientific knowledge constantly evolves as new discoveries appear and old information becomes obsolete. Thus, when the Paleontology Hall Committee first met, it assumed that a 3-billion-year-old fossil would be included as the "oldest" known to science. Before the work was done, however, even older fossils were found and a specimen some 3.8 billion years old was borrowed for the display. While certain substitutions can be made during a hall's lifetime, and new elements added, the planners must settle on a collection of specimens and explanations that will survive both the passing of public fads regarding scientific subjects and changes in scientific mind through new research.

As soon as the scientists start deciding what a hall will be, the Exhibits Office begins planning how. Once a designer draws an initial floor plan with new entrances, exits, walls, ramps, levels, balconies, and so forth, the magical business of turning a concept into a dynamic space starts. Need-

Pages 210–11: *A dozen dioramas in the North American Mammals Hall have fascinated two generations of visitors with examples of the taxidermist's art augmented by habitat models and background murals. Here three elk or wapiti—bull, cow and calf—browse the Yellowstone country after the autumn's first snow, actually 600 pounds of Epsom salts. Now found only in parts of the Rockies, elk once ranged across all North America. They were last seen in Pennsylvania in the 1860s, in Michigan until 1895.*

Pages 212–13: *The puma (a.k.a. cougar, panther and mountain lion) preyed on large hoofed mammals such as deer—and cattle, after European settlement—from Canada to Patagonia. Highly adaptable,* Felis concolor *once thrived in mountain, desert and forest habitats. They are now restricted to scattered pockets of wilderness such as the Everglades and Yellowstone Canyon.*

Pages 214–15: *Three moose graze in the simulated splendor of early autumn. The largest of living deer, they still haunt the Rockies and the Maine woods. With them here, beneath the image of Alaska's Mount McKinley, are a ptarmigan, three-toed woodpecker, Canada jay and red squirrel.*

Pages 216–17: *A mother grizzly teaches her cubs to dig for ground squirrels near Logan Pass in Glacier National Park. Never found east of the Great Plains,* Ursus horribilis *lived mostly on small mammal prey, fish, roots, berries and carrion from the Arctic to Mexico. A few still haunt the Rockies with more inhabiting Canada and Alaska.*

Opposite: Diplodocus longus, *a veteran of the old hall awaiting relocation in the new, surveys the dust and chaos of its surroundings during renovation.*

less to say, the process is positively evolutionary. The placement of doors dictates visitor traffic; the choice of a key specimen the size of a dinosaur alters the shape of the hall. Each proposal by scientist and designer alters the other's options.

The work resembles the outfitting of a luxury liner at one point, at another the drafting of an academic manifesto by the faculty of the University of Babel, and finally the staging of a Wagnerian opera in modern dress with all the loud vocalizing that implies. (The official opening, usually a gala affair for hundreds of invited guests and VIPs, arrives with the frenzy of an opening night. A hundred touches need finishing; nerves fray and tempers flare as the number of hands seems too few for all the tasks. When a recent exhibit required several dozen silk banners and Exhibits technicians found themselves far too overworked, an editor was drafted to iron all the flags so they would hang in wrinkle-free splendor.)

The number of considerations can be mindboggling. For example, the first "unveiling" in the latest Dinosaur Hall occurred nearly a year before the formal one. When the hall had been renovated twenty years earlier, designers decided they'd seen enough of *Diana of the Tides*, a monumental painting commissioned when the building was young. They hid it behind a wall which was removed for the latest renovation. The picture is an allegory, no doubt. One anthropologist wrote a generation ago: "Although primarily intended to represent the story of mythology, *Diana* does in fact picture astronomy, physical geology, petrology, sedimentation, fossils, glaciation, and other subjects of geological inquiry." Admittedly, it shows blue sky and an enlarged shell of the sort that fossil beds sometimes contain, and a patch of the sea which is responsible for many sedimentary rocks. But this 40-foot-wide oddity, hanging high on the wall and suffering a pigmentary affliction called "magenta disease," dominated the room. Everyone agreed it was striking and nobody had the heart to destroy it. The National Museum of American Art wouldn't give it house room and the curators decided in the end they didn't want the curiosity overwhelming their dinosaurs—regardless of its allegorical or antique merit. Finally it was reveiled behind a new wall. The Paleo goddess survives, albeit out of sight again.

Opposite: *Another popular holdover from years past, the stegosaur proved to be made of ordinary papier-mâché—not shredded hundred-dollar bills as tradition had it.*

Left: *A cast of the famous* Archaeopteryx, *owned by the Berlin Museum, mimics the original's colors and texture. In this "hands-on" exhibit, visitors can touch the fossil feathers of the early bird whose toothy jaw and clawed fingers prove its recent descent from terrestrial reptiles.*

Overleaf: *A fossil menagerie greets visitors in the new Dinosaur Hall. Aside from the familiar* Triceratops *glowering at left and the serenely sinuous* Diplodocus *standing aloof at center, two Canadians balance on the ledge along the right wall: the slightly deformed* Albertosaurus *and* Edmontosaurus, *the agile duckbill. Beyond the soaring pterosaur suspended in this enormous room, the metaphorical portrait of* Diana of the Tides *lurks out of sight behind a false rear wall.*

Bony Fish and Sharks

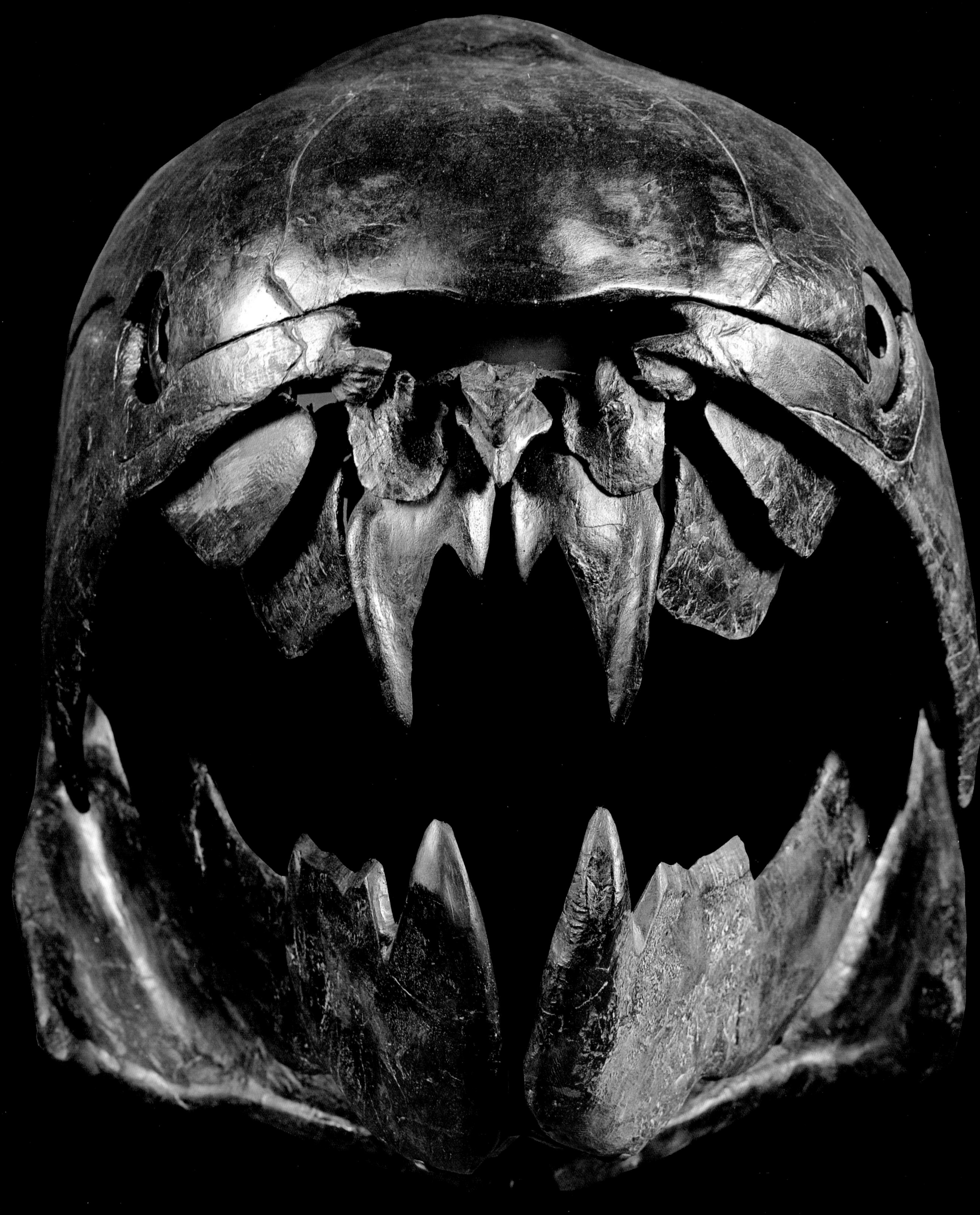

Left: *With a head as big as a medicine ball, heavy armor plates and incisive teeth, the 12-foot long* Dinichthys terrelli *was one of the sea's most awesome predators ever, the tyrannosaur of the Age of Fishes.*

For at least a year before the formal debut, more than a dozen people—scientists, designers and exhibit builders—spend nearly full time on a project. The linchpin is the research associate, usually a postdoctoral scientist hired for the specific job, who comes aboard to serve as expert liaison, arbitrator, interpreter, straw boss and interdisciplinary factotum. This associate works with the committee curators, the designers, builders, contractors and specialists in other institutions to locate specimens, choreograph work schedules and rough-out a "script" containing captions for all the displays.

Tyrannosaurus rex—*actually a new cast of the most notorious dinosaur's fossil skull —receives a technician's cosmetic attention before its debut.*

Each item in the script is subject to a dozen review steps before being typeset for silk-screening on the wall. Members of the ad hoc committee and the full Exhibits Committee all make comments. Revisions ordered by one expert may be revised again by another, then dumped in the lap of a staff writer/editor whose unenviable task is to come up with copy that pleases everybody—or at least offends no one to the point of demanding new changes. Anyone who has ever written technical material for a general audience knows the difficulty of accurately translating arcane idiom into comprehensible English. Anyone who has ever ghostwritten a letter or speech knows the trials of articulating abstractions in another's voice. The research associate must cope with these problems and more. Commanding several scientific disciplines, she must observe all their caveats, including the demand for unimpeachable accuracy, a goal which scientists habitually solve through the insertion of prose-deadening qualifiers. Ghosting for several scientists simultaneously, the writer must sometimes articulate a previously unknown consensus, sometimes describe differences of scientific opinion and always hope for something readable.

While the Museum's craftsmen build many of the display cases and specimen mounts, outside contractors perform the heavy construction work. Sometimes they build cases too and handle special tasks, such as the moving of the 90-foot-long *Diplodocus* to its new location in the hall. In addition, contract artists paint dioramas; and Exhibits Central, a design and fabrication shop serving all Smithsonian branches, takes on one-of-a-kind projects.

The new Paleontology Hall required refurbishing some old fossils and

Building the pterosaur required many skills and stages. First, an artist makes a miniature model out of clay (a). Using a larger, more refined model of wood and Fiberglas, technicians define the full-scale wing plan (b). The wing "bone" is soldered out of metal rods (c) and trussed like an airplane wing—or, as it happens, like modern bird bones—for exceptional lightness and strength. The head is first modeled in clay (d), then cast in its final form and inspected (e). The finished wingbone structure (f) joins the body beneath the suspended working model. Making the translucent wing membrane requires a carefully constructed temporary framework of laths (g), which is then covered with thin clay (h). The clay is grooved (i) so the finished membrane will have spaghetti-like strands of connective tissue along its surface. All this is covered with four layers of plastic resin, which harden to become the flight membrane when the wood and clay form is removed. Nearing completion, the body gets its coat of commercial fur (j)—yes, these reptiles had hair, not feathers. Built in the Smithsonian's Model Shop, the finished reptile is dismantled (k) for shipment to the Museum, a mile away, then reconstructed and suspended (l) ready for its flight high above the Dinosaur Hall balcony.

a

b

c

d

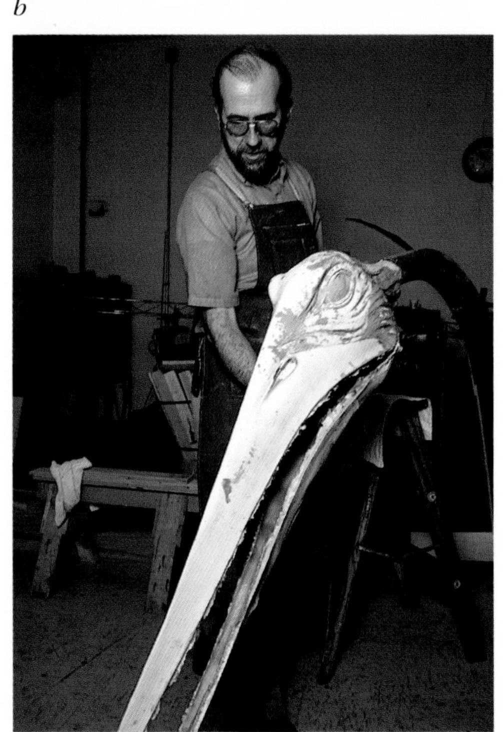

e

f

Overleaf: *The completed life-sized model with a 40-foot wingspan,* Quetzalcoatlus northropi *soars splendidly like his flying reptilian kind of 70 million years ago over the shallow seas that became Texas and northern Mexico. This was the largest of the pterosaurs, or flying reptiles, discovered thus far. The smallest was no bigger than a canary and most of them seem to have lived on fish.*

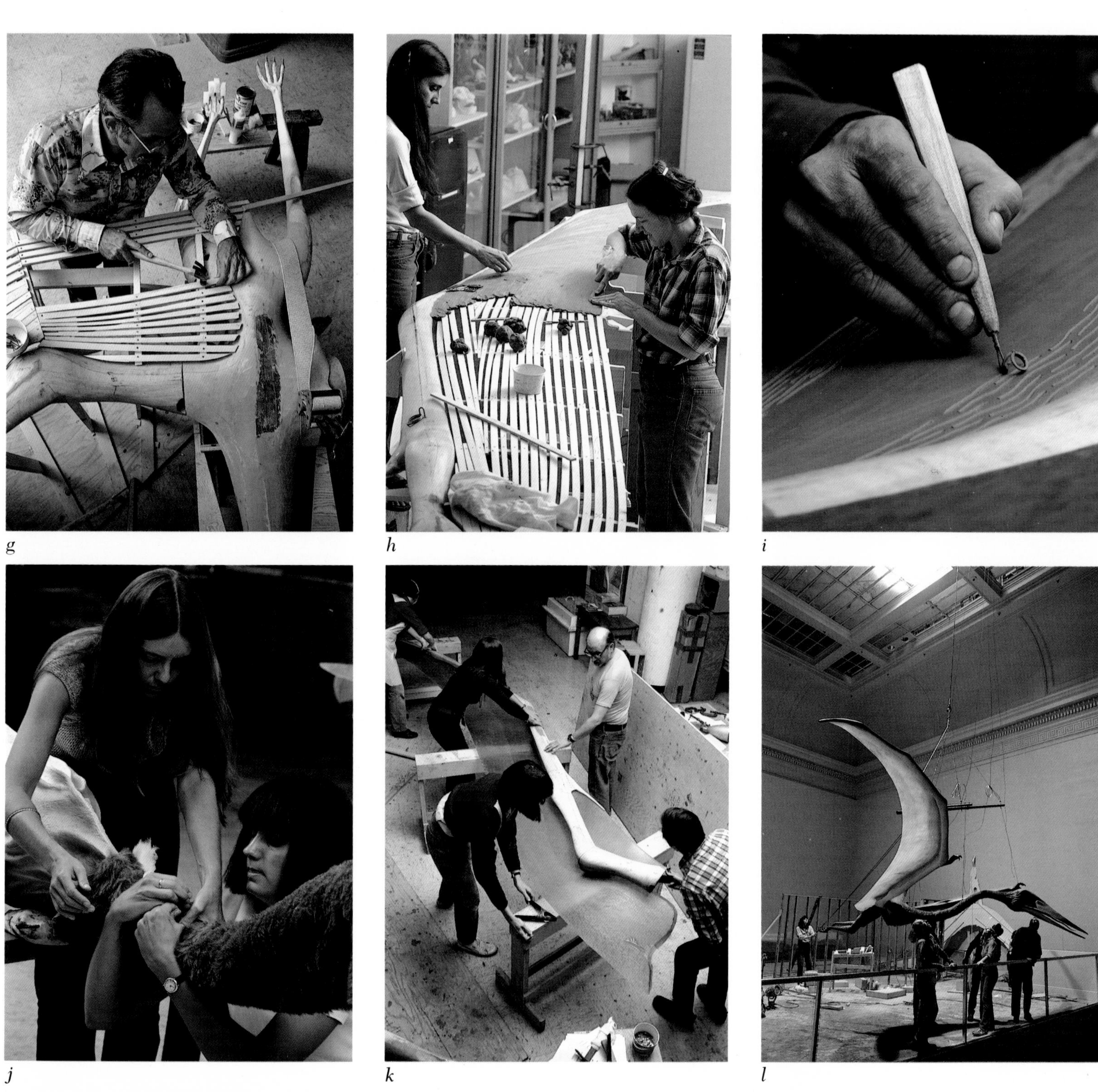

g *h* *i*

j *k* *l*

This sheep-sized dinosaur teaches visitors a lesson in scientific uncertainty. Brachyceratops montanensis (above) *represents a formal species though it may be a baby specimen of* Monolonius nasicornis, (opposite) *a cow-sized relative displayed nearby, which has twice the bulk and slightly different skull features. Were these mammals, their maturity could be deduced from the obliteration of growth zones near the ends of the long bone. In reptiles, however, bones grow for as long as the animal lives and contain no clues to growth stages.*

models, such as the *Stegosaurus* built decades earlier, and creating new ones from scratch. The new design called for placing plants around the lumbering beast. Some living plants closely resemble those found fossilized in the dinosaur's habitat, so a crop of them was freeze-dried, painted, "fixed" with glycerine and installed on camouflaged stems. Other plants associated with the animal are not found alive today, so their fossil fronds were copied in molded plastic. This involved a team of artists and model makers in sketching the plants, planning their fabrication, producing leaves and building a model scene; then perhaps having to do it all again.

Archaeopteryx, a rat-sized reptile bearing wings and feathers, presented other problems. Only a few specimens have been found of this animal which represents the evolutionary link between modern birds and lizards. European museums own the two most famous ones, not the sort of objects likely to be loaned for a 20-year stint. So a cast was required, but quite a special one since the hall's layout called for it to be displayed within visitors' reach. Because curious fingers will wear away any paint, the colors had to be impregnated in the material of the cast itself. It called for considerable knowledge of pigments, plastics and casting techniques plus trial-and-error testing to come up with brown epoxy bones that look like a fossil locked in a synthetic matrix that looks like tan stone.

Exhibits Central's pièce de résistance for this hall is the pterosaur, the spectacular flying reptile with the 40-foot wingspan which hangs in suspended descent over the balcony. The Museum located the young paleontologist who knows the most about this kind of nondinosaur whose fragmented remains have been found in such places as Big Bend National Monument in Texas. A team including the young expert, Smithsonian curator Nicholas Hotton III, research associate Jessica Harrison and the Smithsonian model makers had to contrive a design that faithfully represented what is known about these creatures' anatomy, then flesh out the fossil skeletons. The goal was a life-sized model that seemed to soar as its kind had some 70 million years ago over shallow inland seas in search of fish. (This pterosaur probably pursued fish prey much as terns or egrets do today, although it is not directly related to birds. Their similarities exemplify the phenomenon of "convergent evolution.") Construction

Pages 232–33: *Rivals in miniature, two carnivorous relics face off over the corpse of a model duckbill in this Cretaceous Period diorama which reconstructs an entire habitat. In life 140 million years ago, the warty* Albertosaurus *competed with wrinkle-skinned* Tyrannosaurus *for vegetarian prey amid free ferns and archaic palms, plants resembling their modern relatives.*

Pages 234–35: *Two* Plateosauri *wallow in the model of a Triassic swamp 200 million years ago while the lilliputian* Trilophosaurus *scurries across the mud.*

Opposite: Antrodemus, *armed with three clawed toes, lived before its fellow carnosaur* Tyrannosaurus, *whose tiny forelegs had only two toes. The convincingly fierce-looking skeleton of this* predator *was mounted by a master of the craft, Arnold D. Lewis, longtime head of the Paleo Preparation Lab.*

required a multitude of skills. Wing bones were built of small steel rods trussed like warehouse roof beams, then covered with carefully shaped wood and webbed with a translucent plastic membrane. The head was sculpted, then molded out of polysulfide rubber. The jaws were built first of foam core, balsa and modeling clay. For the eyes, wood hemispheres were lathe-turned, then cast in clear plastic and fitted with paper irises and capillaries of red felt fibers. The claws were carved from maple blocks. The body hair—yes hair, as revealed in one Russian fossil—came from a fabric supplier. And there you have it.

Everything evolves in an exhibit, even the Time Column's lighting. The tower itself grew from a simple enough idea: to represent physically the vastness of biological time. (The committee decided early on to neglect the full expanse of geological time. If the column had represented the age of the earth on the same scale it would have been 230 feet tall, or four times as high as the roof of the enormous room.) Various graphic ideas were set aside in favor of a three-dimensional object. The major elements were easily chosen: a scale marking off the years in 10-million-year intervals; another scale incorporating the named eras and periods back to the Precambrian; the names of successively dominant life forms over the course of 700 million years; finally a stunning macro-illustration showing the evolutionary march of life. The question remained how to illuminate it. Since spotlights always involve glare and shadows, why not let it radiate from within? Since the column would represent a dynamic procession through time, let the light move and give the whole tower some visible activity.

With that charge, the audio-visual staff set to work on what would be the evolution of lights. At first it seemed a string of bulbs wired to go on and off sequentially like an old movie marquee might do the trick. But that meant a heavy electrical load, and the effect didn't look quite right in a model. Then an elevator arrangement was planned to shuttle a bulb-carrying platform up and down. But this meant a heavy physical load and complicated machinery vulnerable to mechanical breakdown. Then why not just haul a reflector up and down inside the column to catch light from a source at the bottom of the tower? In fact, simply painting the bottom of

Opposite: *An electrician works on "the mantelpiece" (also visible on page 223) supporting two dinosaurs that ranged Canada, as their names imply:* Albertosaurus, *a meateater or carnosaur, and the larger* Edmontosaurus, *a big duckbill. As the Jurassic Period ebbed into the Cretaceous 135 million years ago, many vegetarians* (above) *boasted spatulate snouts for the same reason as herbivorous ducks: they work well for grasping grasses. Toothless in front,* Parasaurolophus *(foreground) and* Corythosaurus *had ample molars within their mouths.*

Following pages: Dimetrodon grandis, *a mammal-like reptile that ranged Texas more than 250 million years ago, may have used its dorsal fin to regulate body temperature as the animal faced the sun or offered its profile to warming rays.*

the trolley white would work. This solution meant the trolley need not be heavy—the weight of brackets and a piece of plywood—while the rheostatted lights could remain within the column's base near the motor. If it sounds simple enough, it still required enough adjustments to make Rube Goldberg appear an efficiency expert.

Other design details evolve because all the problems can never be identified in advance. Blueprints showing boxed-in ledges to support two mounted fossils called for straightforward frames built out of two-by-fours. But when a carpenter went to work he realized the framework abutted a case supporting *Dimetrodon grandis*. This old fossil, recovered from the old hall, boasts a fan-shaped array of dorsal fin bones several feet long. One hammer blow to nail the adjacent frame together could shatter the bones, so the plans were scotched and the job was done with screws. (Everyone had already become sensitive to vibrations in this hall since the day a piece of *Diplodocus* fell off. The five-pound spinal process simply tore loose from its vertebral mooring and hit the ground a foot from a senior scientist. No one could explain it until a slight irregular vibration was felt through the platform near the dinosaur. What caused the vibration that came and went? The escalator some distance away perhaps or a bad bearing in an air conditioner, or the city's subway, or a basement machine like Mineral Science's huge rock slicer? The cause was never run to ground. The fact remains: this enormous building shakes from its foundations to the false dome over the Rotunda which, when struck lightly with a fist, rings with a sound like the drum of doom.)

Another recycled fossil, mounted like a ship's half model, was reinstalled in a new wall with bolts and screws that would be hidden from view. Realizing that future carpenters would have to dismantle this case around the turn of the century, the 1980s installers anticipated problems for their successors. To pull it bodily from the wall would risk destroying it, so the installers drew a sketch of their work, complete with locations of bolts and screws, and tacked it inside the wall where future renovators would see it. Instructions from the past.

Installing an exhibit requires considerable hard-won experience. For example, if the ground inside a Plexiglas cabinet is to be covered with

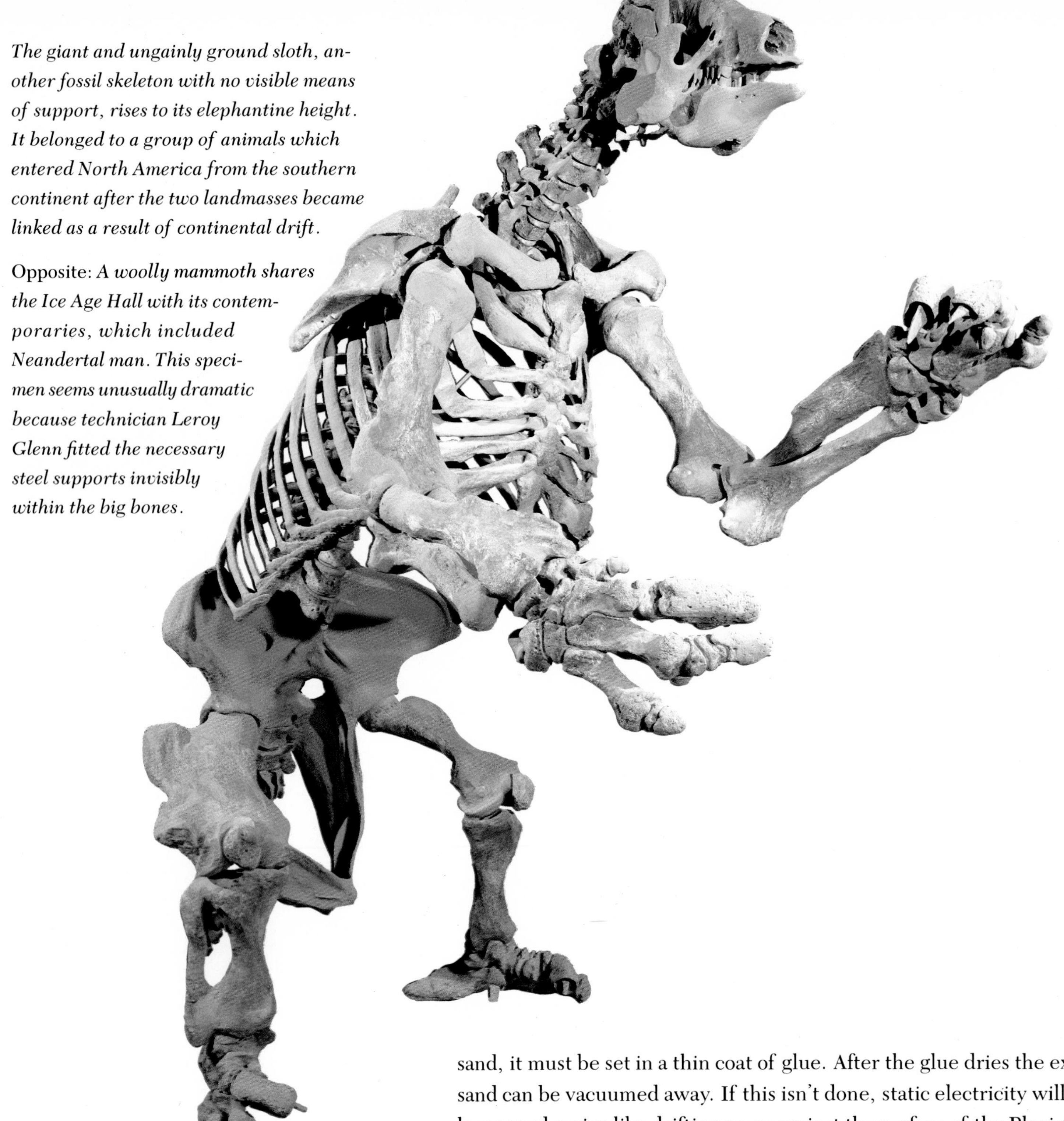

The giant and ungainly ground sloth, another fossil skeleton with no visible means of support, rises to its elephantine height. It belonged to a group of animals which entered North America from the southern continent after the two landmasses became linked as a result of continental drift.

Opposite: *A woolly mammoth shares the Ice Age Hall with its contemporaries, which included Neandertal man. This specimen seems unusually dramatic because technician Leroy Glenn fitted the necessary steel supports invisibly within the big bones.*

sand, it must be set in a thin coat of glue. After the glue dries the excess sand can be vacuumed away. If this isn't done, static electricity will pile up loose sand grains like drifting snow against the surface of the Plexiglas. That would prompt the cleaning crew to open the case to dust the pane, a chore that would inevitably scratch the transparent plastic and ruin it. Thus, for long-lasting success, proper installation of a show involves surprising savvy and attention to countless details—many of them invisible unless they go unattended.

In other instances, a display can require a shipwright's skill, a Chinese-puzzle maker's ingenuity and the patience of Job. For example, 35 man-years of work went into the reconstruction of the *Diplodocus* skeleton. Even then the task required the help of a nearby Navy shipyard, which forged a serpentine steel member about 50 feet long to support the dinosaur's spine. In the nearby Ice Age Hall stands the woolly mammoth

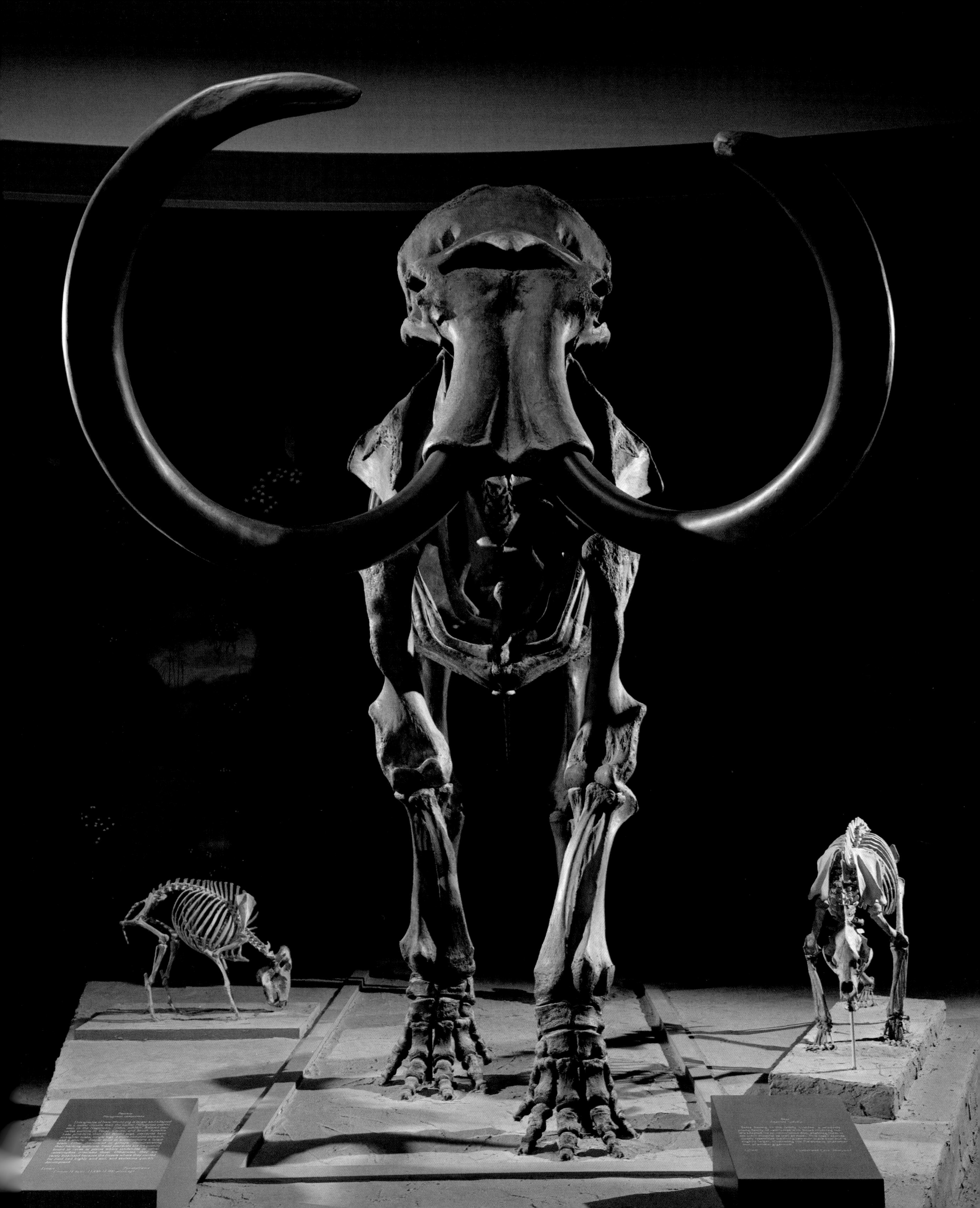

Harvesting

that a technician spent two years piecing together. It seems a self-supporting relic standing on its own four feet because that technician bored through the length of massive long bones to construct, piece by piece, a steel skeleton within the fossil one. The giant sloth across the hall also rears up in perpetuity thanks to hidden steel supports for the brittle bones.

Other exhibits employ outright illusions, the stuff that delighted magic show and theater audiences in the nineteenth century. The prime example is the "Pepper's Ghost" box, which allows two objects to mysteriously come and go, replacing each other in one display case. The device employs reflective glass. When a viewer looks through one pane at illuminated objects behind it, he sees what's there. When that interior light goes off and another goes on outside the glass, he sees whatever's reflected off the angled plane, namely another carefully placed object. John Henry Pepper, a chemistry professor at England's Royal Polytechnic Institute and a popularizer of science, patented the device in 1863. In his time it was used to make ghostly apparitions "appear" on stage before awestruck theater audiences.

Pepper's Ghost boxes project the relative sizes of arthropods—from insects to a giant King Crab—in the Insect Zoo, literally the liveliest part of the Public Museum. This loud and very popular room, well named in terms of its captive fauna, is also a menagerie of captivated visitors. Designed with children especially in mind, it features living specimens of all sorts of insects and related small creatures: bees busily maintaining their indoor hive (which has access to the outdoors), a ferocious and poisonous centipede nearly a foot long (in a jimmy-proof case) and enough tarantulas for public feedings every day (though each eats no more than one cricket a week).

One purpose of this room is to introduce visitors to the extraordinary variety of arthropods, an ancient group which now accounts for eight out of ten living organisms on earth. (As one entomologist puts it: "Age of man? We live in the age of insects.") Two full-time professionals manage this room, aided by scores of regular volunteers who introduce visitors to their many-legged charges' surprising traits: the docility of the tarantula, the lightning speed of the praying mantis, the days-long copulatory

Opposite, top: *Visitors at the Insect Zoo meet the hickory horned devil, a harmless vegetarian despite its name. After metamorphosis this common eastern American caterpillar emerges from the cocoon as a yellow-and-brown regal moth with a six-inch wingspan.*

Middle: *A potential entomologist studies the behavioral trait which allies* Apis mellifera *with* Homo sapiens, *namely the honey bee's communal habits. Countless other insects, including upwards of 3,500 bee species in North America, collect pollen and nectar to make food for their young. But almost all are solitary nesters. In America only this one species, an exotic imported by early European colonists, gathers en masse to make honey in hives, like one maintained in the Insect Zoo.*

Bottom: *A toddler in the Discovery Room makes the acquaintance of a crocodile rendered harmless by taxidermy.*

embrace of the walking sticks, the nearly silky smoothness of a giant cockroach's back.

More hands-on activity occurs in the Naturalist's Center downstairs, a study facility with a wide range of specimens that sponsors events like the "Draw In" for apprentice scientific illustrators. The Discovery Room also offers firsthand experience in interdisciplinary matters, though many of its clients can hardly pronounce such polysyllables. They are schoolchildren and families who come in groups several times each week to handle real fossils, manipulate such things as "two-part shells" and "shells that spiral" (bivalves and sea snails), and puzzle over common archaeological artifacts.

Like the Insect Zoo, the Discovery Room could hardly stay open without the active work of volunteers, some of the nearly 1,000 people from junior high school boys to widowed grandmothers. Some man the information booths, others catalogue minerals, several gifted youngsters spend summers preparing microscope slides for the Entomology Department. (At least two eminent curators, one a geologist, the other an insect expert, got their starts as teenage volunteers here.) Yes, the Public Museum—which includes the building's restricted inner sancta—owes some of its continuing activity to members of the public who find their way through its inner doors. The National Museum of Natural History not only educates and entertains the public, it gains much of its vitality from the ranks of ordinary folk who offer both their stimulating curiosity and their energetic services.

Opposite: *Outside the Museum, children clamber on a life-sized Fiberglas model of* Triceratops, *a.k.a. "Mr. Beaseley" after the television character who was born from an egg found by a fictional farm boy.*

THE MUSEUM OF ART

Dedicated to Science, the Museum displays countless treasures of man-made beauty. The recently installed Thomas M. Evans Gallery on the ground floor, a showcase of temporary and traveling exhibits, most often offers what are intrinsically art shows: "5,000 Years of Korean Art"; a matchless collection of Kachina dolls; the priceless Alaskan Eskimo artifacts of the Nelson Collection. One-man contemporary shows of natural history watercolorists and photographers often hang on the Rotunda's third floor. But the art throughout this building is not always so public nor so apparent.

Perhaps a camera never lies, but in many scientific circumstances it cannot see the comprehensive truth, let alone capture it. Many things of special interest to investigators from anthropologists to zoologists can only be illustrated by artistic hands: fossil mammals wearing their flesh and hides again, the detailed portrait of a tiny moth or gnat, the faded pattern of a pot broken in the throes of an ancient culture's disappearance. All these kinds of objects, and many more, require the special artistry of scientific illustrators, a community that has its center here.

Objects like these elude photography for a variety of obvious and obscure reasons. Insects, for example, are notoriously three-dimensional things while a microscope lens, captive to the laws of optics, has a stingily narrow depth of field. When an insect's carapace comes into focus, its head goes fuzzy and the legs pass out of sight. A microphotograph can rarely capture the whole animal in clear detail. Even if it could, each individual has abnormal features and each specimen is distorted to some degree by dust particles or the suspension medium. A skilled scientific illustrator can compensate for these optical distortions, compare several specimens to discover typical characters, and constantly focus up and down to comprehend the entire animal. Every research scientist can do these things, of course; they're central elements in the repertoire of microscope technique. But most scientists don't draw very well. The illustrator adds an artist's indefinable ability to translate three-dimensional objects onto a single plane and preserve the image on an easily reproduced sheet of paper or plastic film. It hardly comes as a surprise then that the pen-and-

Prized by the Museum for its anthropological value, this mask remains an eloquent work of art. Carved by a nameless St. Lawrence Islander and collected by Captain Hooper of the Revenue Cutter Corwin *in 1881, it is unusual in its depiction of an old man, the sort of elder whom Eskimos admired for his worldly wisdom and experience. The artist's motive can only be surmised, but the previous winter two-thirds of the island's people died—from famine or imported disease, perhaps—and the 400 survivors found themselves in dire circumstances in the Bering Strait. It is just one of many forms of art found in the Museum.*

Overleaf: *A camera's narrow depth-of-field defines the nearly spherical compound eyes but blurs the rest of the insect. This specimen represents 400 species of "robber flies," so called for their predatory habits. Among the fastest flying insects, they catch live prey on the wing.*

Botanical artist Candy Feller pursues her work in submarine situ *off Central America using a drawing board buoyed up by air chambers and held down by anchors.*

Opposite: *George Venable illustrated this tiger beetle using a carbon dust technique, a medium that permits precise detail and rich contrast.*

ink drawings which illuminate the intentionally colorless prose of scientific papers often possess extraordinary beauty.

The Museum has employed illustrators for more than a century, but it is only in recent years that they've begun to receive recognition commensurate with their contributions to science. Traditionally they received no credit at all, says senior artist George Venable. Now that's changing, thanks in part to the growing Guild of Natural Science Illustrators. Quite clearly biological literature has grown to depend on the patience, experience and talent of trained illustrators. For example, as an authority on water beetles compiles the definitive monograph about these insects in North America, he must describe upwards of 1,300 species. Many within a genus closely resemble each other in every respect but one: the male genitalia differ. (In fact each is so different that the male of one species cannot physically engage the female of an otherwise identical species any more than a radio bought in Lima, Ohio, can plug into a wall socket in Lima, Peru. It's a case of physical equipment enforcing the "reproductive isolation" that by definition determines a species.) As a writer I am compelled to aver that these differences of form can be described in words alone—but at what a price. Relying only on written descriptions to identify a specimen in hand would try even an entomologist's extraordinary patience. Convention requires detailed written description, but illustrations make the distinctions immediately apparent.

The flexibility of art enables things to be depicted that don't occur simultaneously in life: for example, all forms of a metamorphic animal such as a crab, which may experience a dozen different shapes during its life. Other organisms change drastically when transferred from their natural habitats; jellyfish and marine plants don't even look the same when preserved in solutions. But a selective drawing can freeze their vital appearance.

The craft of marine illustration took a sizable step forward recently when the Museum helped a freelance illustrator visit Belize and join investigators studying the mangrove swamps. An experienced diver, Candy Feller had developed an underwater drawing system complete with what amounts to an aquatic easel. Its key part, a drawing board fitted with

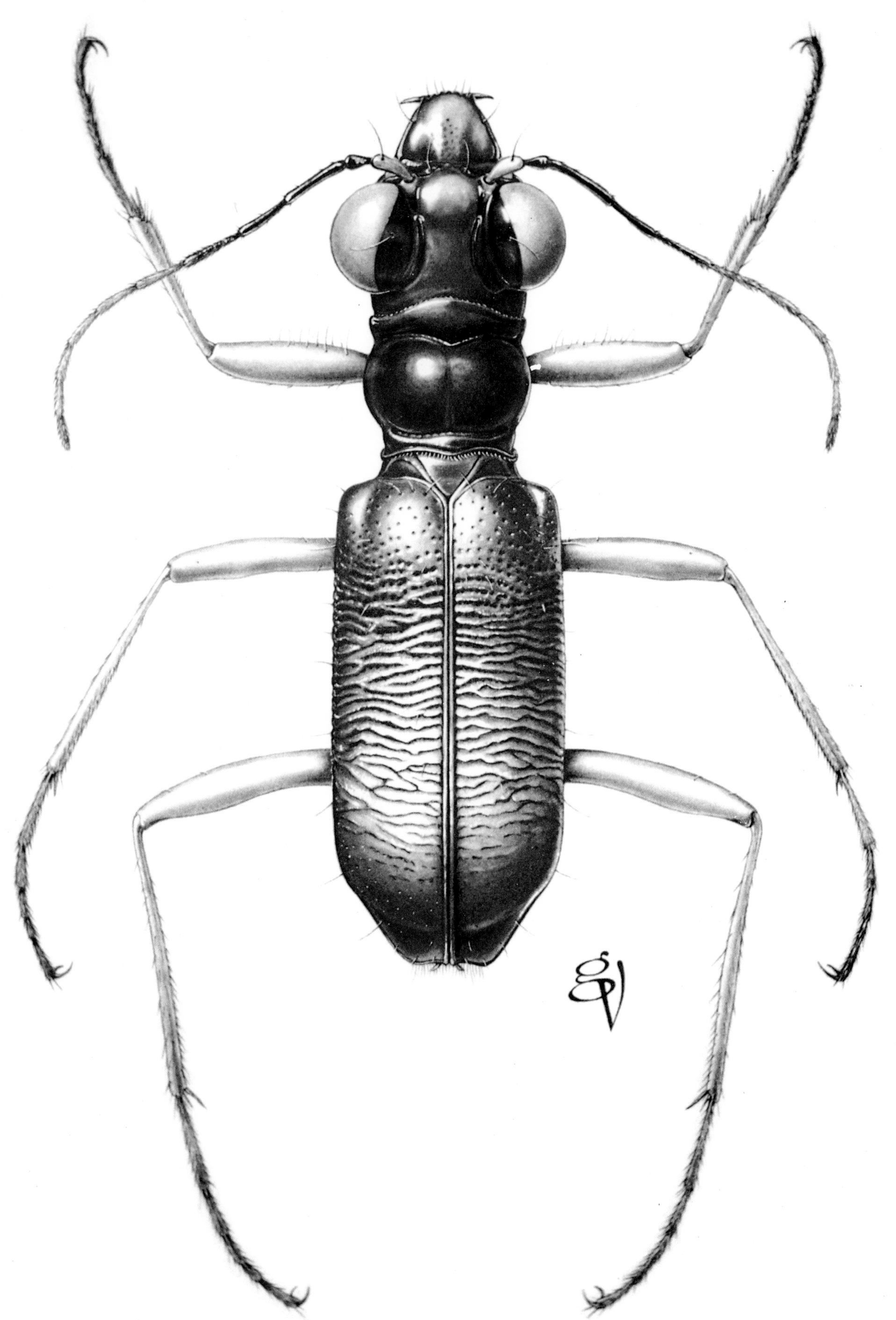

Above: *This ink stipple drawing by George Robert Lewis of a Southeast Asian stone mask was one of a series of illustrations for a publication by Dr. Betty J. Meggers suggesting possible ancient trans-Pacific migrations of pottery-making seafarers to coastal Ecuador.*

Right: *This mixed media drawing by Lawrence B. Isham, who specializes in paleontological material, details the cross section of a clam to reveal how various structures and tissues relate.*

Opposite: *Alice Tangerini's pen-and-ink drawing illustrates almost everything a botanist needs to know about the anatomy of a newly described wild Brazilian legume,* Brodriguesia santosii. *The central figure, called the "habit sketch," includes leaf structures and inflorescence. Various vital structures surround this (clockwise from lower left): the pod, seed, leaflet gland, petiole, flower, bud and flower in sections.*

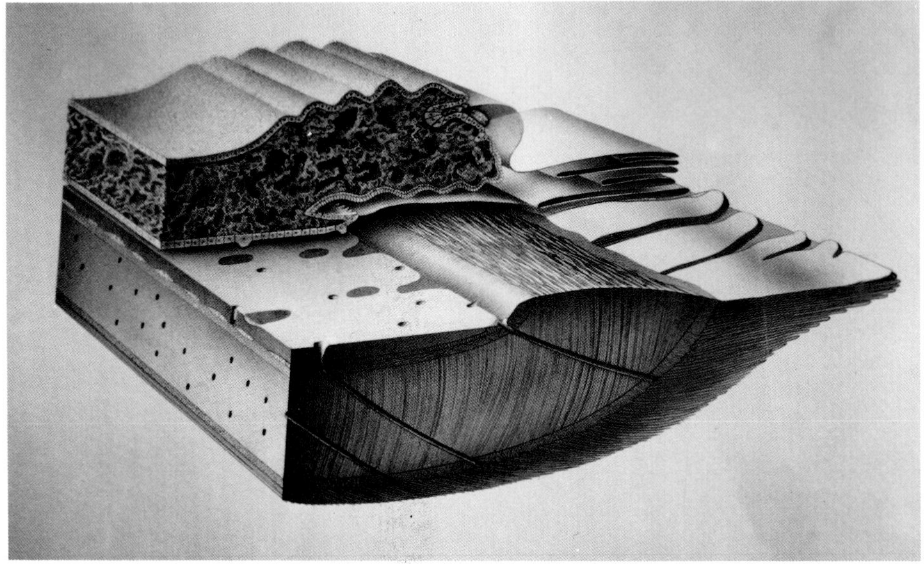

a 2-inch air chamber, is held down by 24 pounds of diver's weights and buoyed up by the trapped air to become a surface as stable as many desks. Instead of soluble paper Feller uses a polyester film and hard pencil. For future color exhibits in the Museum she also tested a color chart which allows her to determine the precise apparent color of an organism under water, then replicate it on dry land. In the past artists had to guess or remember an organism's hue, a chancy and subjective process since the colors change at different water depths.

Anthropology also employs illustrators for a variety of tasks. The Department's working group often makes "roll outs" to reveal the design running all the way around a piece of pottery or other circular artifact. Single-viewpoint rendering or photography can't do this. Neither can a camera capture barely visible incised patterns nor uniformly reconstruct a painted design long since faded or worn away save for dim flecks of color. As for broken specimens, film can capture pieces of trash well enough, but the artist can take shards and draw a seamless reconstruction of the pot as it once appeared without waiting for a technician to piece it all together again.

The same group also makes innumerable maps and charts of archaeological digs. The maps, sometimes adapted from a conventional atlas or

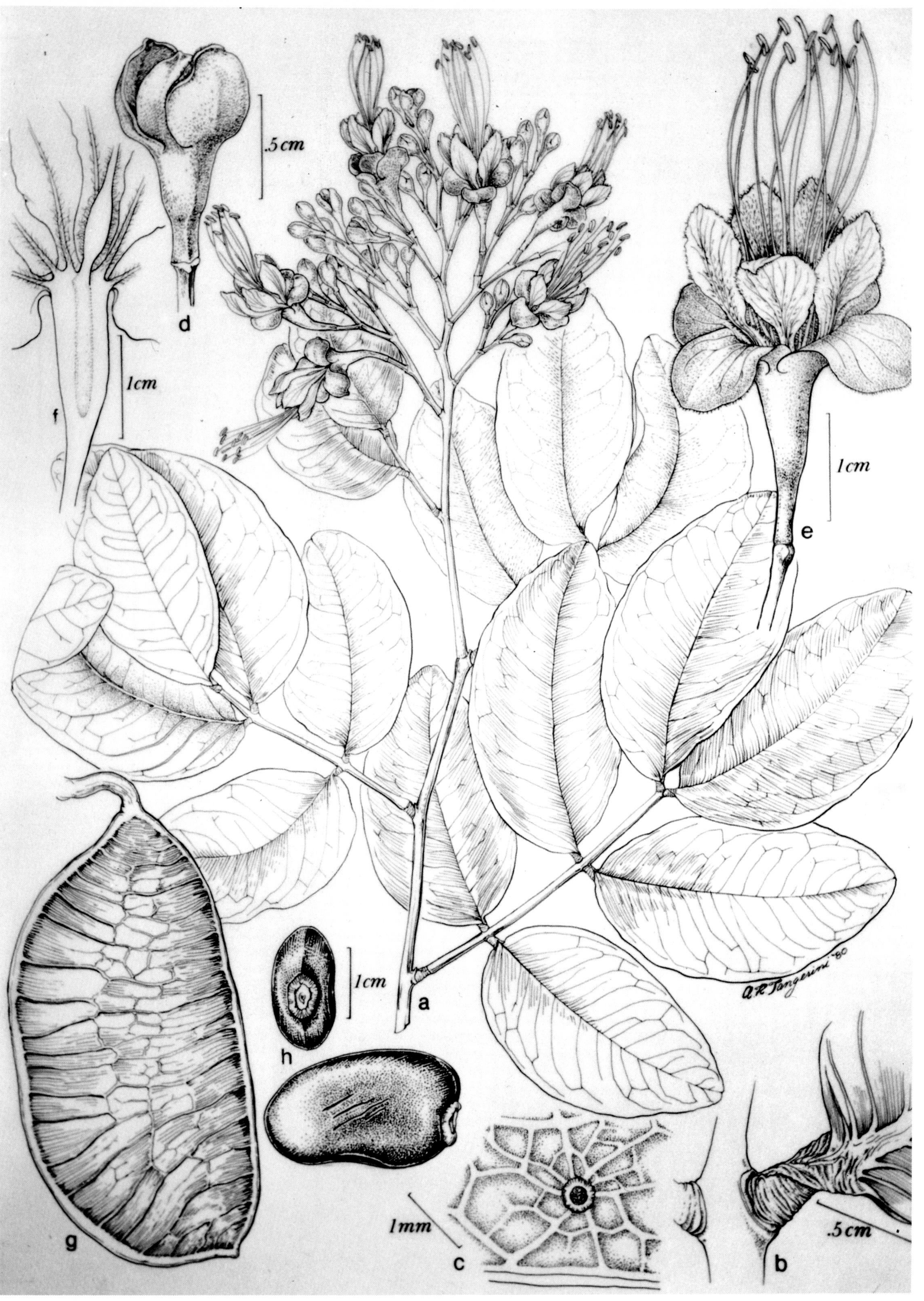

.5cm
d
1cm
f
1cm
e
1cm
h
a
A.R. Tangerini '80
g
1mm
c
.5cm
b

land survey, can highlight important physical features and ignore extraneous ones. Drawings of a dig as it progresses can record the location of bones and artifacts as layer after layer is removed.

Discovering what can never be seen again—indeed what was never seen by human eyes—involves another branch of scientific illustration. Paleontologists around the world regard Jay Matternes as an artist of rare biological genius. Indeed he is both a gifted painter and an experienced anatomist. Early in his career he painted several large murals depicting "the age of mammals" for the Museum. His technique involved reconstructing the musculature of these animals from their fossil remains: the shape of bones, the size and location of "muscle scars" where the ligaments attached, the outline of vanished soft tissues. He extrapolated stature from often-fragmentary remains of several specimens. He deduced the dynamics of an animal's motion from the articulation of ball-and-socket joints. Finally, after studying all the geological and fossil evidence pertaining to a place and period, he could add local flora and landscape to compose a portrait of an entire habitat. One important element remained for his discretion: the color of the animals' coats, since the fossil record contains no information about pigments.

More recently Matternes has pioneered the accurate reconstruction—rather than the fanciful exaggeration—of early human faces from skeletal remains. Working with the Museum's dean of physical anthropology, he portrayed a Neandertal skull for a popular scientific magazine. Starting with the Shanidar skull restored and studied by T. Dale Stewart, he proceeded through a series of conceptual overlays using the same kind of technique he'd employed to put the flesh back on the extinct brontotherium with its double nose horn and the trilophodon with its shovel tusk. Layer by layer he added muscle to the bony foundation, then fat and skin, to discover that our primitive Neandertal ancestor had a very familiar look about him.

In the interim a singular artist from the Commonwealth of Virginia Medical Examiner's office approached a Museum osteologist in search of new (or as it happened very old) material. This law officer reconstructs recent forensic specimens, i.e., the faces of people who died mysteriously,

Pages 256–57: *Murals can offer hypothetically rich and varied arrays of contemporaneous animals. This one depicts a possible scene during the Eocene Epoch (50 to 45 million years ago). Another advantage of artistry: a series of murals can highlight the zoological history of a region, as masterpieces by scientific artist Jay Matternes show on these pages.*

Pages 258–59: *This one depicts an unlikely moment in the Oligocene Epoch (37.5 to 22.5 million years ago) somewhere in present-day South Dakota.*

Opposite: *During the Miocene Epoch (22.5 to 5 million years ago) grasses spread over the earth and pushed the forests back. Animals adapted and evolved in turn.*

Pages 262–63: *In the Pliocene Epoch (5 million to 1.8 million years ago) mammals continued to diversify across the drier Great Plains of North America.*

Pages 264–65: *The Early Pleistocene, less than 1.8 million years ago, knew many birds and small mammals which survive today. The "Great American Faunal Interchange"—result of animals crossing the newly formed isthmus—was at its peak, bringing South American species like sloths, porcupines and possums north to this continent.*

usually outdoors, and whose partial remains must be positively identified. If most of the skull and facial bones survive the work of foraging animals, a great deal can be done to restore the face because the thickness of muscle layers is fairly constant from individual to individual. (Little flesh lies over the temples, more over the cheekbones, still more along the jaw.) Using modeling clay to mimic muscle strands, he builds up the face into an often recognizable likeness which can be crucial to a police investigation. To provide him with diverse examples, the Museum came up with some ethnically unusual skulls which were copied as plastic casts. These skulls had been unearthed from Bronze Age burial chambers in Jordan.

Yet another Museum scientist, Dr. J. Lawrence Angel, plans a chronological series of hominid skulls reconstructed as busts for a future exhibit on human evolution. The coincidence of three scientists all using sculptural reconstruction represents an interesting example of what one of them calls "mental convergence." As an architect can look at a building's steel frame and visualize the finished skyscraper, each of these anthropologists can look at a skull and accurately imagine the face it once presented. (Long experience enables Angel to specify age, sex, race and occupation from a piece-by-piece look at a skeleton. He then backs up his intuitive opinion with detailed analysis of bone shapes and relationships. Men tend to have heavier brow ridges than women, Negroes broader nasal openings than Caucasians, children less complete skull closure than adults, etc.) But most laymen can't read much in the bare bones.

Conventional art techniques as old as the paintings in the Lascaux caves show up all over the Museum, especially in the Recent Mammal Hall dioramas and various anthropology displays. The muralist's delicate stroke-by-brushstroke approach is also augmented by silk-screened silhouettes, diagrams and captions on the walls.

In this era of computer art, "found art," etc., it only stretches the definition of artistic media slightly to include scanning electron microscope photography. The SEM produces beautiful pictures by almost anyone's definition: the sharply sculptured shapes of pollen grains, the toes of insects resembling lunar vehicles, the confectionary balls of diatoms, the tree-like annual rings of a dolphin's tooth. Atomically arcane, the SEM's

Above: *Scientific illustrator Anne E. Lacy, whose credits include the illustrations in several popular books and a forthcoming monograph on North American beetles, fleshes out livestock on a mural in the Western Civilization Hall. The finished work provides a vivid backdrop for the sculpted denizens of Al-Kosh* (opposite).

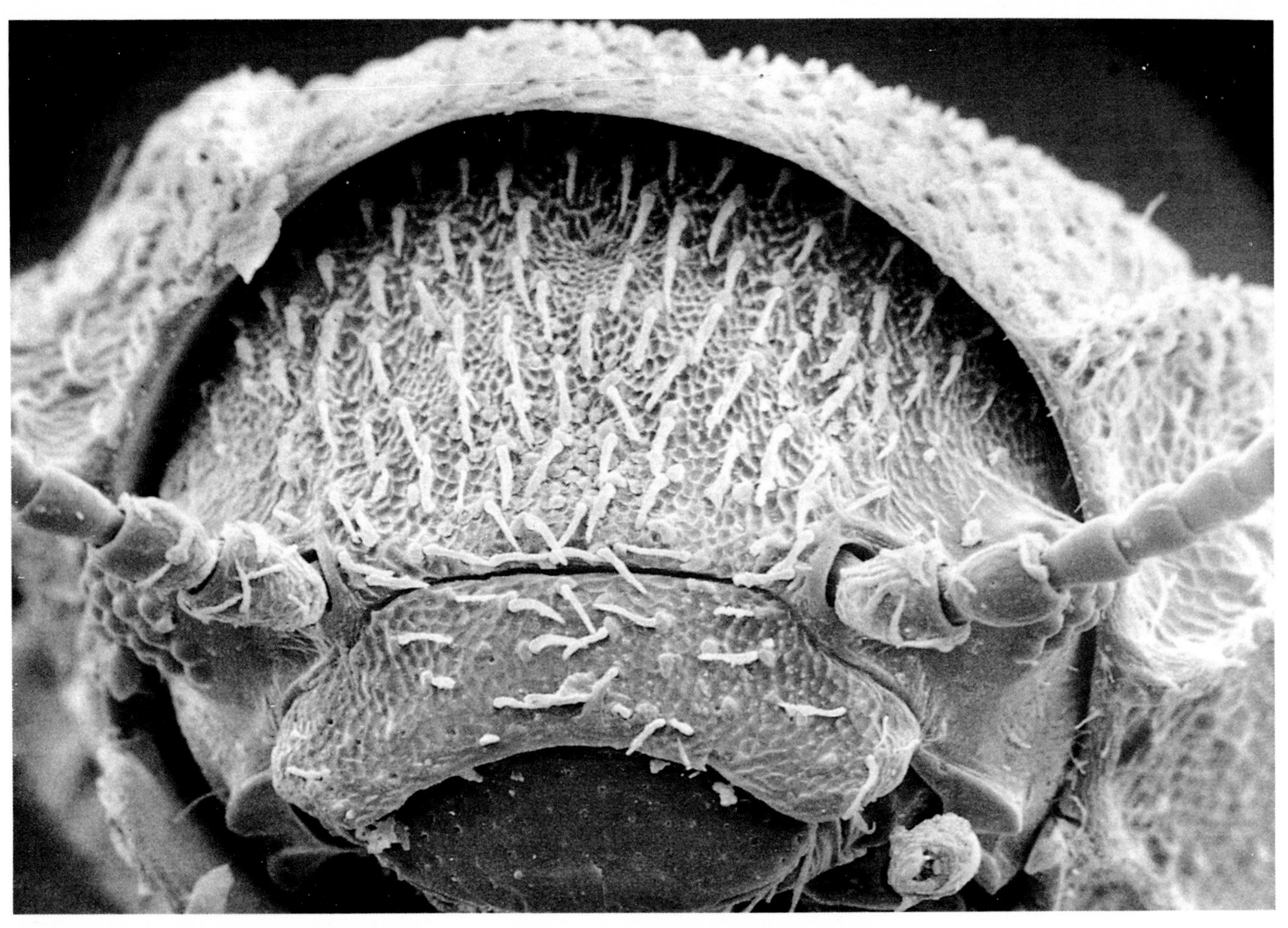

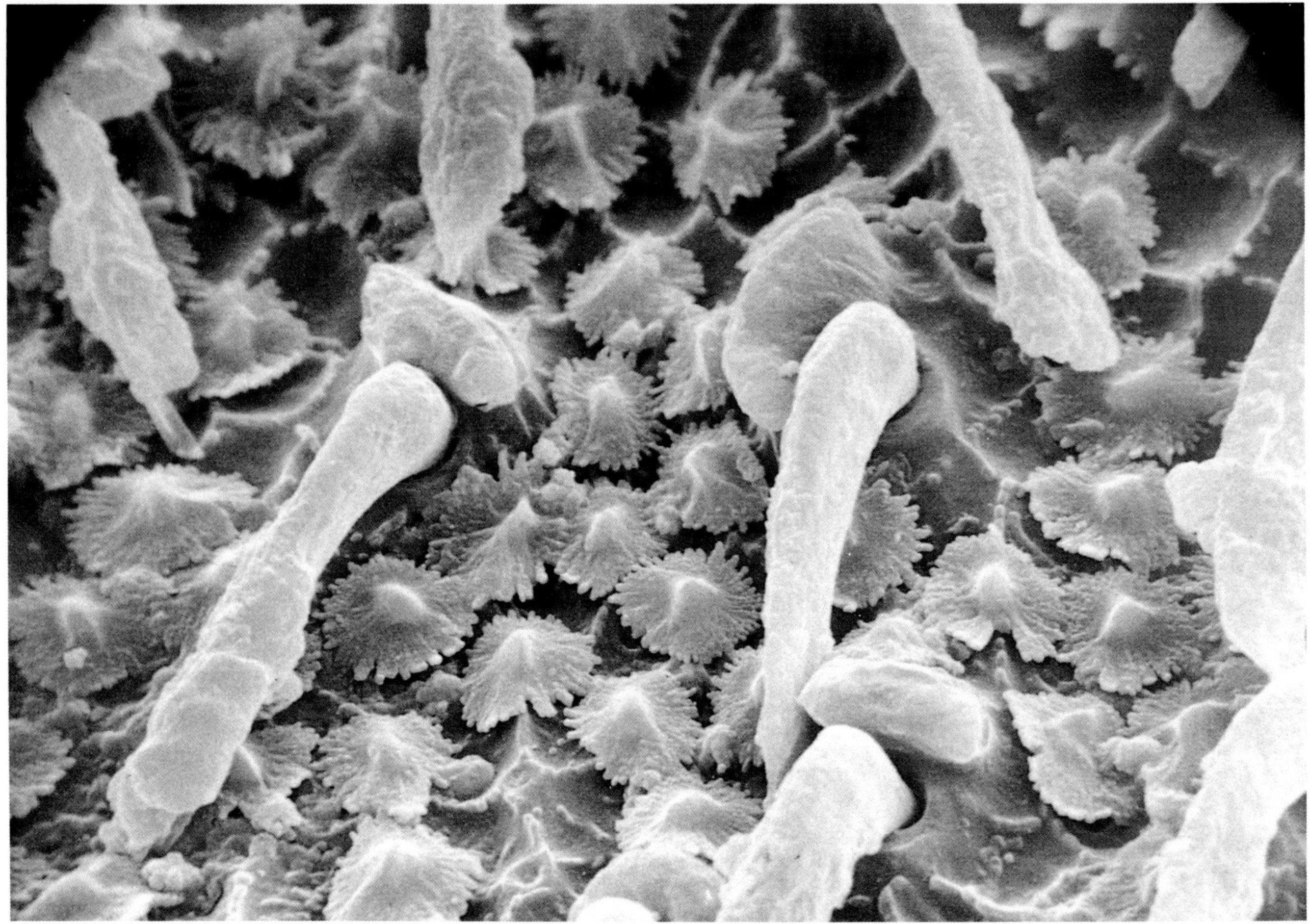

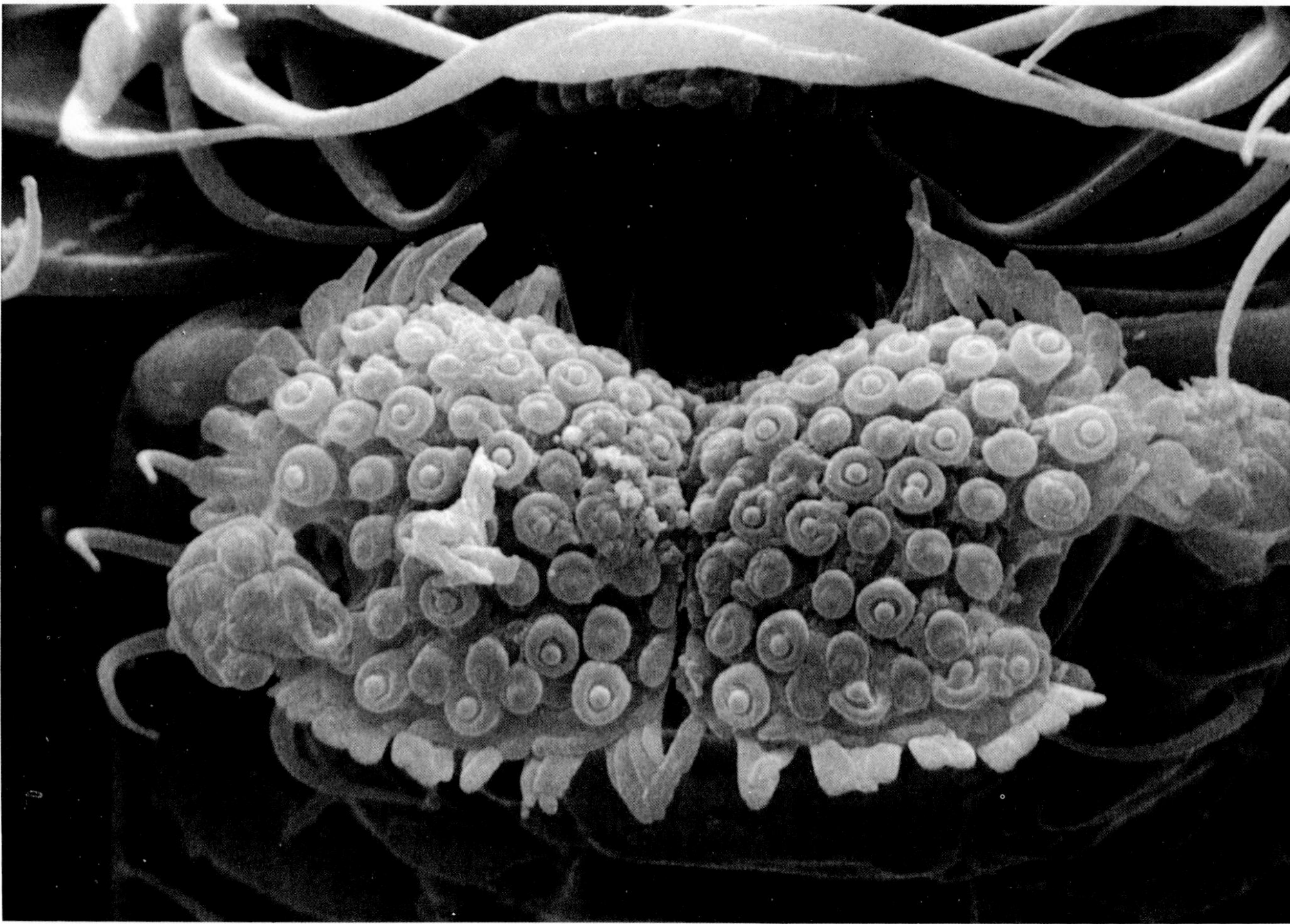

Preceding pages: *The Scanning Electron Microscope (SEM) produces remarkable images of enormous magnification. The head of the Haitian water beetle,* Lemalelmis fontana *(left top), was magnified 185 times on the SEM (then about thrice again for publication); the hair-like setae (left bottom) between its antennae, 1,500 times. When the head of another aquatic beetle in the genus* Sphaerius *(right top), magnified 430 times, displayed an odd lip, the Lab magnified it 2,550 times for closer study. Anatomical structures—"characters" in the trade—are as important to insect taxonomists as a deer's antlers and sheep's horns to a mammalogist. But before SEM's advent, this* Sphaerius *could barely be seen since the whole animal measures about .5 mm, or the size of a typewriter's period.*

Opposite: *Pollen grains of three different members of the geranium family, all magnified more than 2,000 times, display their diagnostic architecture. Those most alike within a family are usually most closely related. Magnification more than 7,000 times* (right) *reveals even greater detail.*

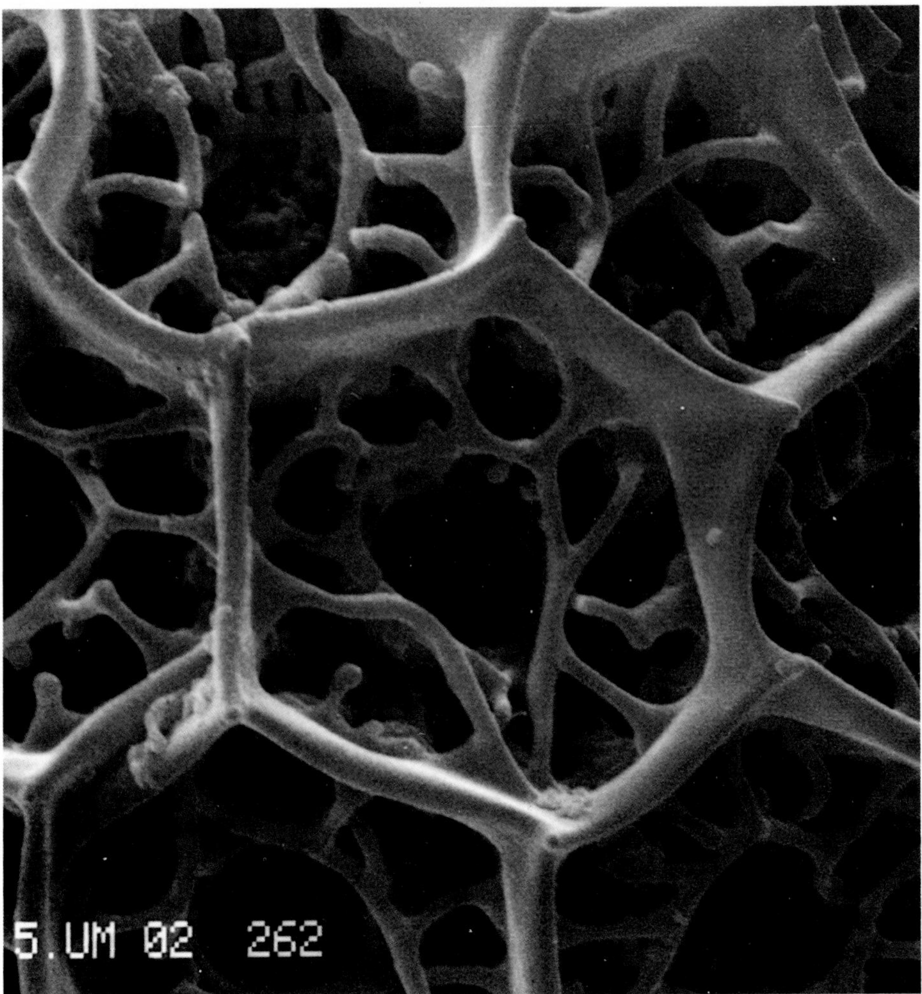

gross mechanics are relatively easy to comprehend: a specially prepared specimen is placed in "a fairly good vacuum," approximating the atmosphere 100 miles up, to remove obstructing air. A beam of electrons is focused through a lens while magnets scan the beam back and forth across the specimen. Some of the electrons bounce away erratically; more of them bury themselves in the sample, dislodging electrons from the specimen. Flying away with characteristic strength, these are collected by a positive electrical field that attracts the negatively charged particles. The minute electric signal thus produced is amplified to produce an image on a television screen. Because the number of freed electrons varies with the angle at which the beam strikes each part of the specimen, a useful picture emerges: flat areas appear dark, slanting areas brighter, steep planes brightest. In the process the specimen can usefully be enlarged as many as 30,000 times.

Obviously a laboratory that employs four people full time must produce something more than pretty pictures. Corals expert Frederick M. Bayer explains it succinctly. The colonial animals he has studied for decades have characteristic spicules—minute skeletal structures of calcium carbonate scattered through the gelatinous organic tissue of the coral colony. More definitively than a coral's gross anatomy, the shapes of these tiny structures are diagnostic as to the organism's "species, genus, family and right on up the line." Thirty years ago, Dr. Bayer—one of the

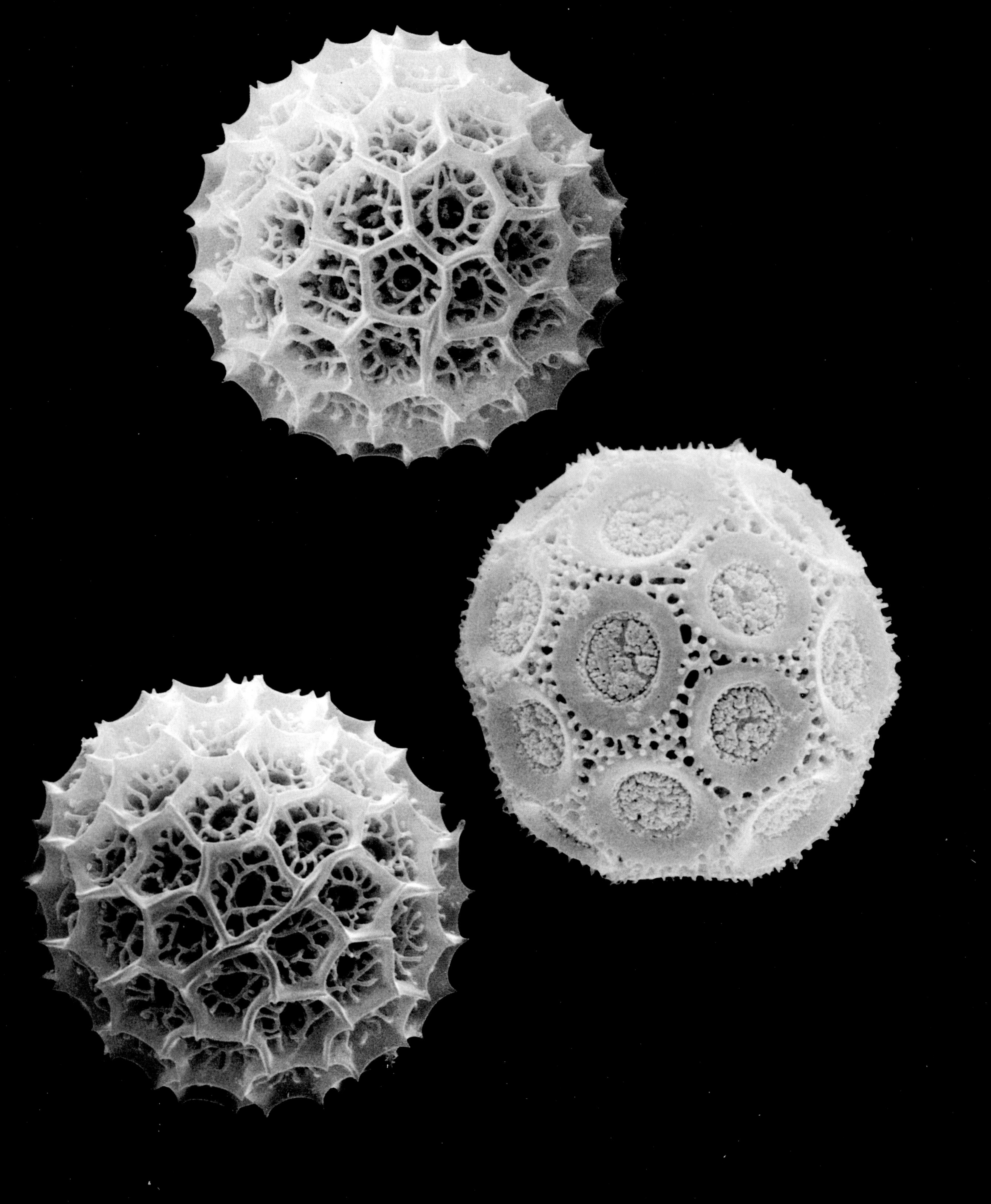

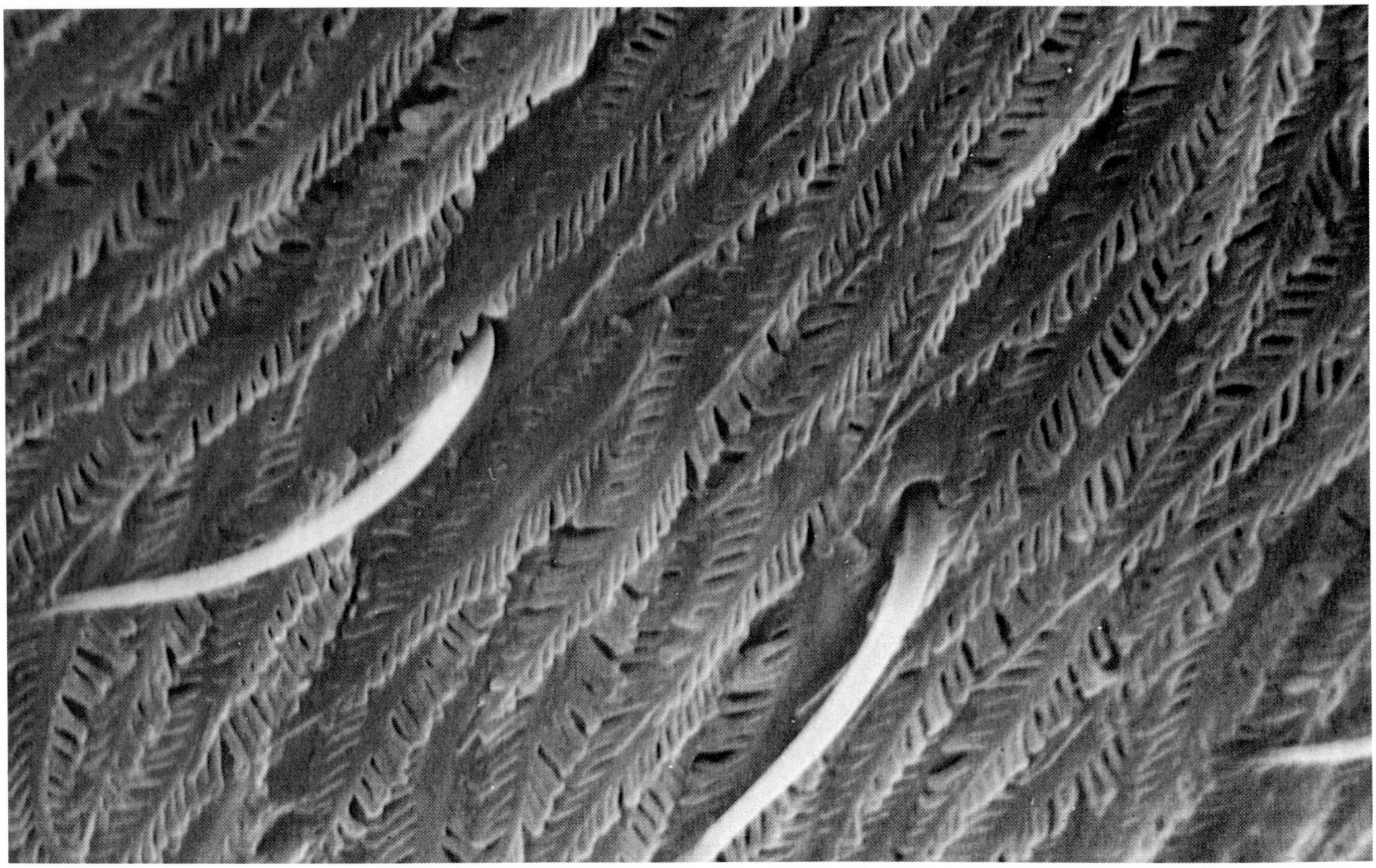

scientists who *can* draw—would spend a week at a stretch hunched over a light microscope to draw the various diagnostic spicules of a single species. "This afternoon we've just done twelve in 20 minutes," he said during one of his regular stints in the SEM lab. Not only can the machine produce more perfectly accurate photographic images faster than a human can draw, but individual images at high magnification can be pieced together in "mosaics" to reveal the microcrystalline structure of a large surface area, graphically revealing the way the spicule was formed by the coral.

One disappointing aspect of the SEM: it only functions in monochrome (since its minute specimens have characters with shorter dimensions than the wavelengths of visible light). But this limitation had no importance when an entomologist examined birdwing butterflies famous for their iridescent colors. Their wing scales are colorless; the spectacular decorations are not pigments but "physical colors"—the result of white light's refraction through a prism (whether crystal or organic). One controlling factor for each color is the vacant space between arched layers of transparent cells that make up the wing scale like the brick columns, walls and ceiling of an old wine cellar. Distinct patterns were seen in scale "architecture" as if, say, one butterfly's scales were laid up cell by cell like bricks set in Flemish bond and another's out of adobe.

Just as a European mason and a Navajo can build square houses, birdwing butterflies of different phylogenetic heritages can flash identical colors although the cell arrangements in their wing scales differ markedly. Conversely, very slight differences in one arrangement or another refract white light differently and produce distinct colors. In the past taxonomic relationships were often assigned in part on the basis of color similarities. It now appears that the genetically determined architecture has more phylogenetic importance than the superficial effect. Thus the scale structure holds a new key to discovering relationships among different groups of these Southeast Asian insects. The invisible and involuntary art of cellular composition, revealed by the color-blind SEM, provides a new tool for determining the taxonomy of a scale's brilliantly colored possessor. Chalk up one more surprising and practical instance of visual beauty in the scientific Museum.

Opposite, top: *The setae on this aquatic beetle's body capture an exquisitely thin film of air, which allows* Austrolimnius formosus *to breathe underwater, unlike most species, which must regularly come to the surface. The air film itself acts as a kind of respirator, absorbing dissolved oxygen from stream water and releasing carbon dioxide.*

Opposite, bottom: *Starfish-like, these setae belong to a Haitian beetle that inhabits deep colonial wells. Recently discovered by Paul J. Spangler, it is the only known eyeless member of its entire family.*

A MUSEUM OF EARTH

As the goddess of wisdom Athena leapt fullborn from Zeus' brow, the Museum of Natural History grew so basically out of geology that the two reasons almost flout straight-faced explanation. First of course, James Smithson made his mark as a geologist; his most lasting contribution to knowledge involved minerals. Second, without earth itself, the seminal focus of geology, there could be no other natural science.

Geology began with the investigation of earth's accessible parts, then reached progressively both higher and deeper. In the process, the single science spawned such diverse disciplines as crystallography to examine earth's minerals and astrophysics to explore the larger context in which earth exists. If interdisciplinary expansion slowed down in recent years, geology as a whole advanced by mundane and extraterrestrial bounds. Intramurally the revolutionary concept of plate tectonics provided an organizing principle as important to understanding earth itself as the idea of evolution had been to the birth of comprehensive biology. Externally the space age opened the door to new insights into the celestial realm and earth's place within it. Through all of this the Museum has been both an abiding resource and an active contributor. Its collections, containing astonishingly ancient evidence of the planet's past, provided new answers to old questions, for instance: Why do mountains still rise? And why do volcanoes erupt near earthquake zones? Armed with new tools, Museum scientists (and colleagues in specialized institutions around the world) have written the planet's history in previously unimaginable detail. Importantly, proof of many events always lay within the collections here; it only took new keys to unlock newly seen compartments of data in the mute stone. The upshot is that the Mineral Science Department's specimens have lately revealed new details of the physical world's biography. And what a story it now appears to be.

All things being relative, "old as the hills" turns out to mean something younger than the Man in the Moon and thereby starts a tale that the Rock and Mineral Halls on the third floor tell in mind-stretching detail. Written in stone, the script is being translated from molecular cipher by geologists, "historians of the earth," in the Department chairman's phrase. These far-flung investigators freely borrow material from the nearly half-

Crystals of smithsonite, the zinc carbonate first recognized by James Smithson, cover a base rock found in Namibia. By definition a mineral crystal is a solid with a specific composition and internal structure. As one grows, its component atoms interlock three-dimensionally to create a characteristic shape—cube, pyramid, rhombohedron, etc.—derived from its internal geometry. Many substances solidify this way; indeed in technical terms all inorganic matter occurs in only three forms: gaseous, fluid and crystalline, i.e., solid. A single substance can acquire different crystal forms under different conditions. For example, pure carbon commonly appears as graphite, the softest of known crystals, but also forms diamonds, the hardest, when subjected to extraordinary pressure.

Overleaf: *Cracks in rock (or mud) are often filled with later material that may become harder than the original. Here a quartz honeycomb, over a foot long, is all that remains to show the cracks in a septarian concretion that has weathered away.*

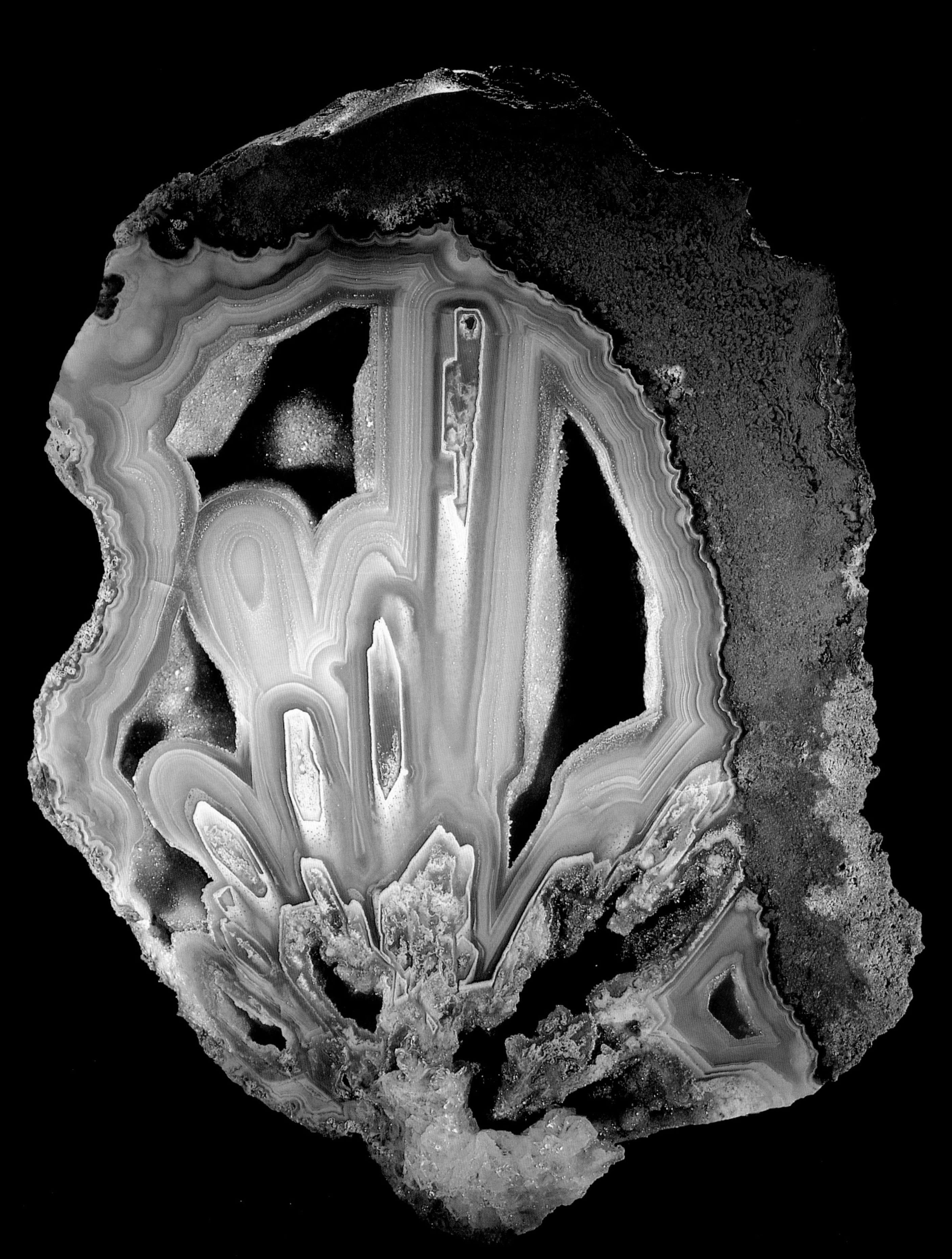

million specimens found in the exhibits, ranges and strong rooms of the Museum's East Wing. Without this collection of meteorites, rocks, minerals and gems—one of the two or three most comprehensive in the world—today's geologists would be like scholars without source books. They would be blind readers in lightless libraries without the remarkable tools and techniques at their disposal here and elsewhere: electron microprobes, radioactive dating, X-ray diffractometry; older implements like brute machines that break rock in an instant or compress it in hours, precise saws that slice it paper-thin in the course of months, perfectly balanced horizontal grindstones called laps that polish specimens with diamond dust and abrasives softer than rouge. Many of these tools are new while many a specimen, preserved since the Museum's founding, has an even older provenance. This dichotomy points up the inestimable value of the curatorial museum per se: an object that presented old conundrums can be examined again in new light and new objects compared with old ones under identical conditions.

Stones are the geologist's texts, and if a history book can be said to have vitality, so the poetic metaphor "living rock" has a basis in fact because this most solid stuff undergoes slow but constant change. A Scottish pioneer-naturalist named James Hutton suggested the importance of this phenomenon to earth science about the time young Smithson came into his fortune two centuries ago. As described in the Physical Geology Hall nearest the Rotunda, Hutton offered the then revolutionary hypothesis which has since won the mantle of demonstrable geological law. In a nutshell: the processes that made earth what it is continue; the world's physical history did not depend on sudden, once-only events but on continuing phenomena that are timeless to all human intents and purposes.

Applications of Hutton's Theory of Uniformitarianism appear simple enough. To understand the present world, study clues found in its own ancient records; conversely to visualize the past, observe earth's dynamic processes as they occur today. Why? Because silt still settles mote by mote in Chesapeake Bay, slowly compacting into layers of potential sedimentary rock. Because Mount St. Helens' latest explosion brought a minute fraction of earth's molten innards to the surface, where it became new rock

Opposite: *Agate, one variety of silicon dioxide, forms in geothermal regions. This complex specimen, found in a Mexican mine, filled a void in the surrounding rock as water containing dissolved silica passed through. The solution left layer upon layer of crystal material on the walls of the shrinking space until it nearly closed.*

and soil—as rock and soil have been formed over millions of millennia. Because every continent's lowlands have been flooded and drained repeatedly changing their stony constituents. The rocks formed under ages-long "floods," which ebbed as both land level and the sea level fluctuated, commonly contain the remnants of marine animals even after other events thrust old ocean bottoms up into mountains. Such rock is termed sedimentary, one of classical geology's three general categories.

The Museum has samples of them all and each tells of different events. Sedimentary rocks can reveal the environment and biota of a sea during different periods, as well as suggest conditions on the surrounding land. Igneous rocks have fiery origins, like the new dacite on Mount St. Helens' slopes today. This cold stone can reveal its temperature at birth and thus the conditions of the earth's interior whence it came as well as the rates at which it cooled to crystallize into diagnostic minerals. Metamorphic rocks, once igneous or sedimentary, are those that recrystallized when heat and/or pressure from new sources remade the original material. A slice of gneiss near Hutton's portrait demonstrates this: thin, continuous black and white layers curve and recurve like hairpin turns on a roadmap. Each thin layer of color was formed sequentially as a flat, horizontal sheet of sediment, which then crystallized. Later the whole stack crystallized again into a monolithic stone when uneven pressures shaped the laminations like potter's clay.

Rock with plastic properties? In a manner of speaking, yes, given the right circumstances and sufficient time. Page by page, the geologists' systematic studies of earth history reveal a far more complex past than any mythology or science fiction ever depicted. Indeed many aspects of this story defy superficial credibility because our frame of reference is too small. For example, it challenges the intellect to imagine that the marble Michelangelo carved was once base limestone. Similarly it might seem unlikely that a few million years hence today's Mississippi muck may become the stuff of cemetery markers and roofing tiles—if people still bury their dead and roof their homes, indeed if people still inhabit earth. On first hearing it defies common sense that every landmass could erode in a geological instant until the oceans cover all if mountains stopped

This gneiss probably crystallized originally as granite from molten magma deep in the earth. Over millions of years the original rock, changed by heat and pressure and tectonic movement, acquired a segregated (banded) and convoluted crystal structure as metamorphic rock. The gneiss has the same chemical composition as its granite parent, but a different arrangement of its constituent minerals.

A "modern sculpture" of ancient crystals—red elbaite, milky quartz and ropey albite—exemplifies the aggregation of related and distinct minerals under natural circumstances.

Following pages: *In a region where hot springs well up from inside the earth, sulfur escapes in gaseous form. When it cools, say in a subterranean cave, it condenses into nearly perfect crystals, the largest of these about 2½ inches across. A crystal will continue to grow in this way as long as the necessary elements collect and space remains.*

Right: *Shaped like a stellar fish with an eye of troilite mineral, this slice of the Willamette meteorite comes from the largest object of its kind found in the western hemisphere. Called "noble in size" after its discovery by an Oregon woodcutter in 1902, it weighed 13 tons. The granular pattern indicates reheating and secondary crystallization after its original formation in its parent body.*

Opposite: *The "stony iron" Salta (Argentina) meteorite contains crystallized metal and lumps of olivine in an amalgam that could not have formed on earth.*

Pages 288–89: *This "nickel-iron," fairly typical of one meteorite in twenty, fell near Alvord, Iowa. Cut with a band saw and etched with acid, its face displays a Widmanstätten pattern, which is a reflection of the crystalline structure of the component metals. The thickness of the lines is a function of the composition and the rate at which the material cooled within a huge mass over millions of years.*

Pages 290–91: *Another polished and acid-etched section, from a meteorite that fell near Breenan, Kansas, displays its own Widmanstätten pattern. Photographer's lights impart the colors, which have no diagnostic meaning.*

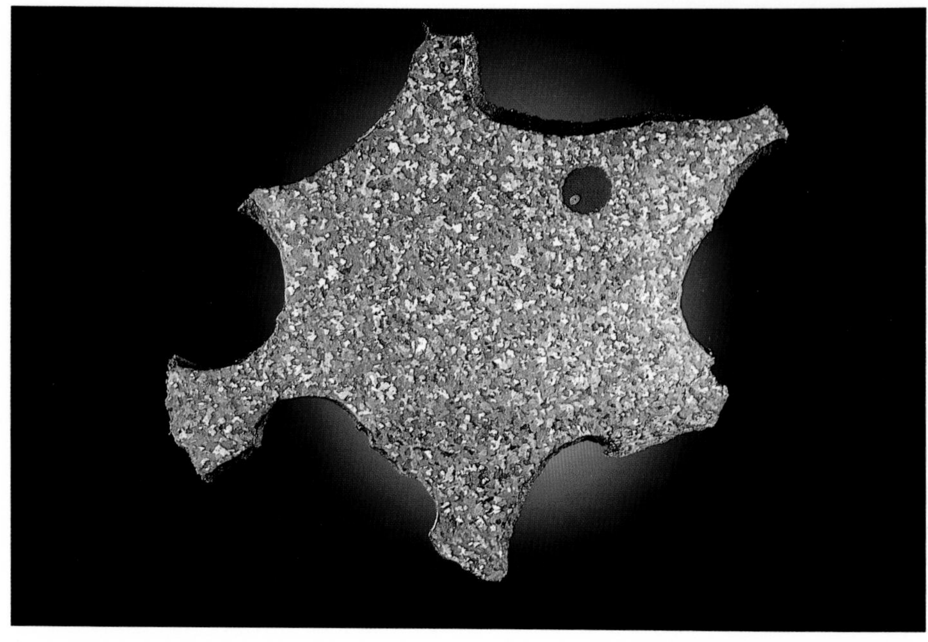

growing. (The nearby Potomac River carries so much sediment into Chesapeake Bay each year that without compensating geological action, Washington and the entire watershed would erode to near sea level within 10 million years.) It's startling to believe a label beside a pint-sized basalt block: it says the stone could explode with the power of 70 pounds of TNT and blow the hall away—if all its original radioactive energy were released at once. And who could have thought in Smithson's day that earthquakes and volcanoes share common origins. The long-term dynamics of the planet, compiled under the heading of plate tectonics (that is, "building"), seem outlandish on first encounter. Be that as it may, earth's entire crust is composed of distinct segments, plates that float on an interior layer, variously colliding, separating and slipping over, under or past each other as they move.

To accept these facts one must abandon human scales of reference for the moment. The enormity of mundane forces and the oceans of time in which they operate involves magnitudes beyond our living experience. To reach a better perspective, start with what we experience as terra firma—a misnomer if ever there was one—and consider that all we see of earth comprises less than 1 percent of the planet's material. Its irregular crust is relatively no thicker than the skin of an apple. The continents seem massive and permanent to us, because by comparison we are minute and

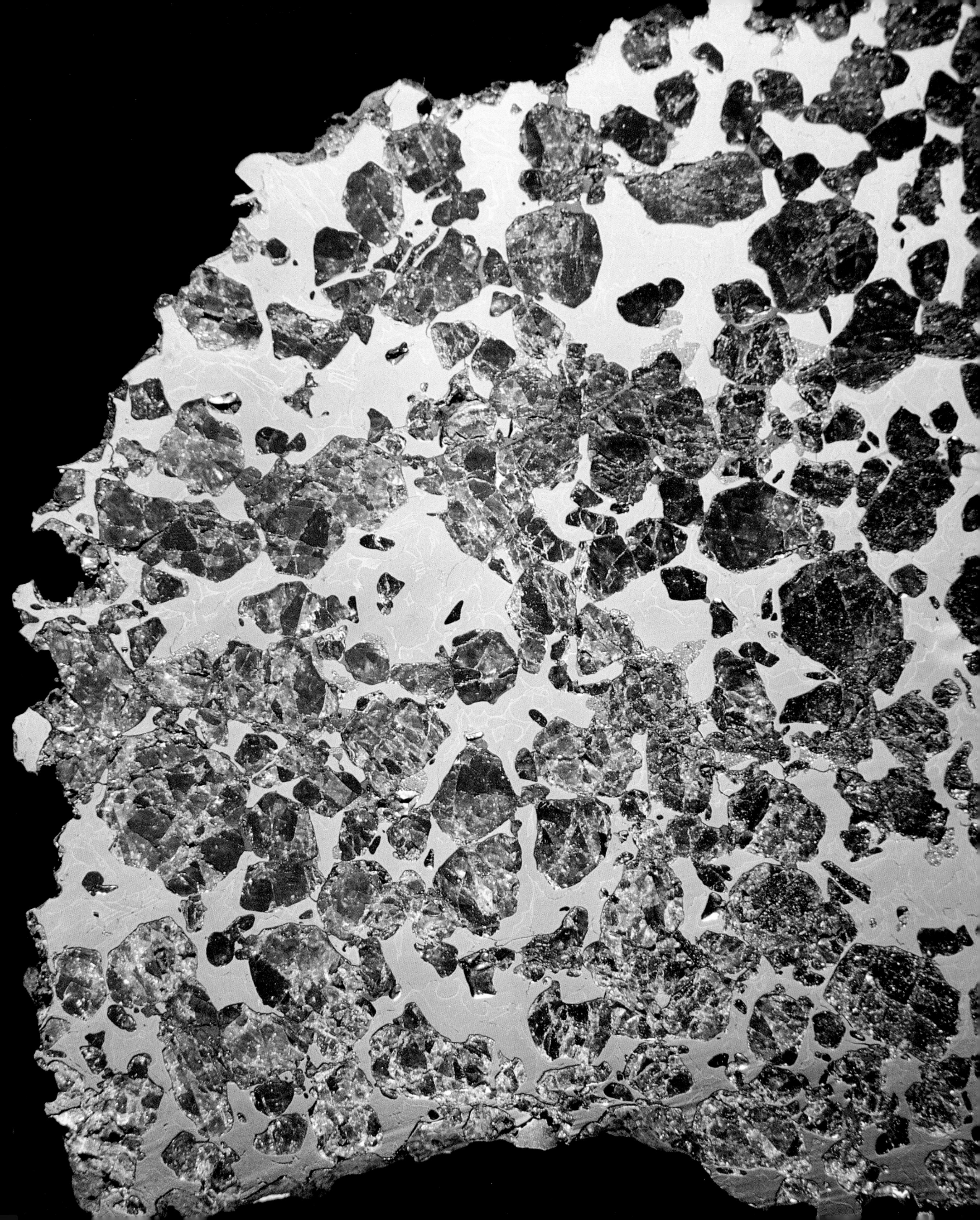

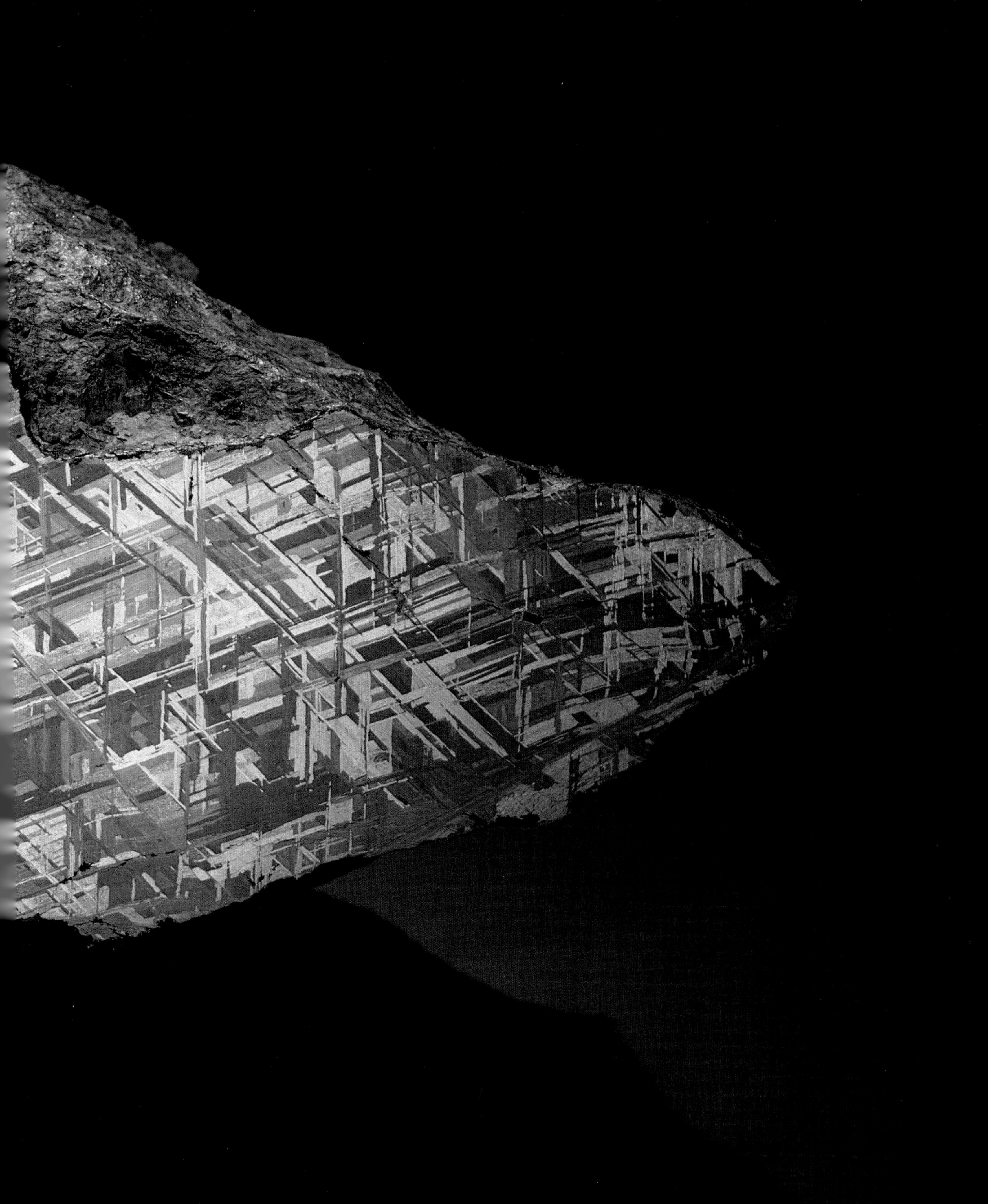

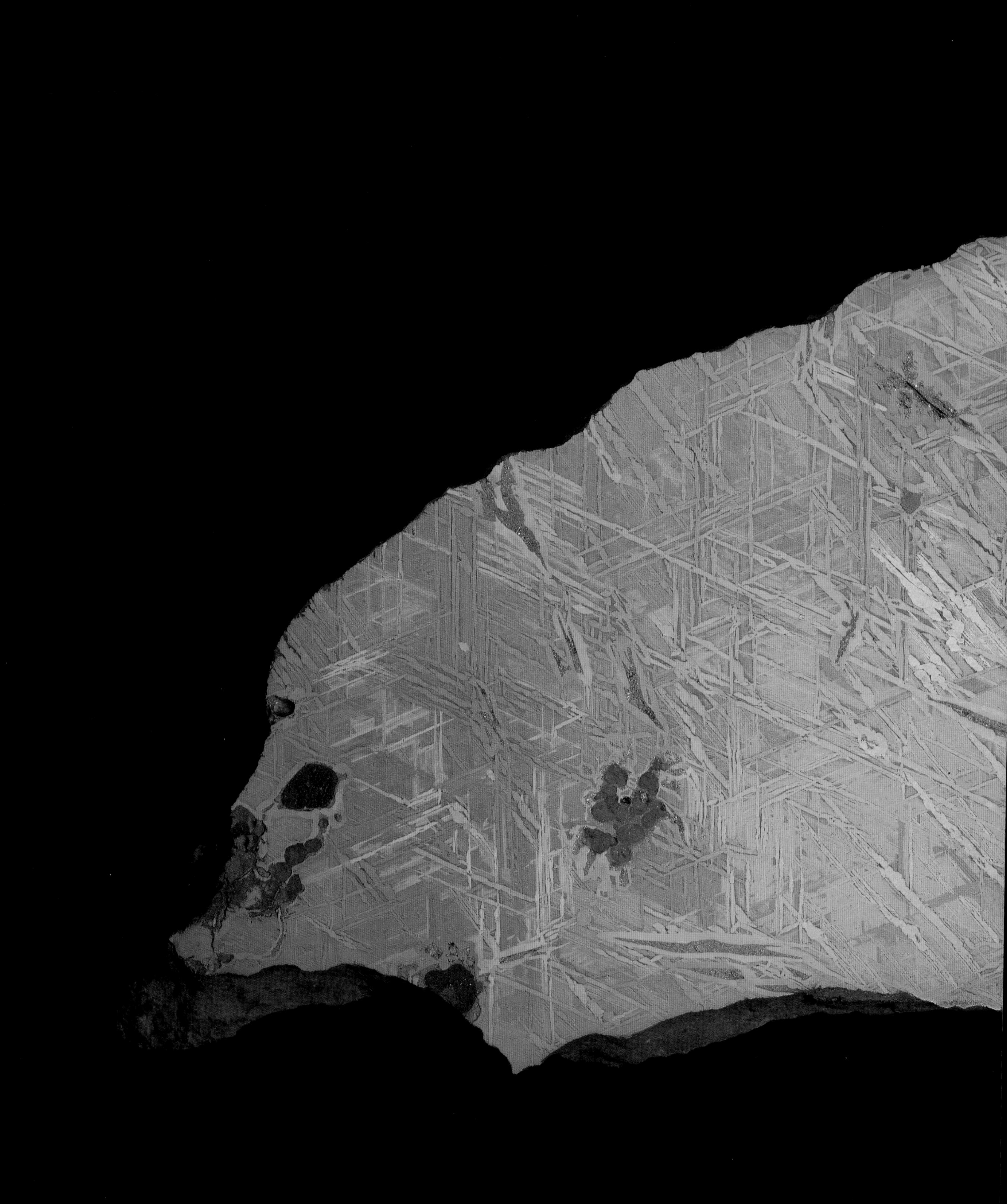

Opposite, top: *This agate, possibly from a petrified forest in Oregon, may have formed in the cavity left by a buried tree that decomposed so completely that no trace of its cell structure remains.*

Opposite, bottom: *Quartz crystals, both amethyst and clear, mingle together. The colored parts owe their hues to nearly unmeasurable amounts of iron locked in the crystal structure—a few parts per million at most.*

short-lived. (Bacteria residing on an elephant's back might do as well as geologists to comprehend the elephant—except that germs never developed a scientific method to test their perceptions, so far as we can tell.) In normal human terms most geological events are imperceptible; yet if the ancient sun were the observer, terrestrial development might appear to occur as fast as a child grows in a parent's eyes.

To grasp an outline of earth's biography in this its middle age, assume new scales for the physical elements and for the time involved, because geology relates intimately with vast time. The solar system and earth began on the order of 4,600,000,000 years ago—that's 4.6 billion, a number as functionally incomprehensible as the annual increase in the national debt. To give it a more familiar context, consider the accepted fact that during the average man's lifetime North America and Africa drift farther apart by a distance equal to that average man's height. North America alone presently drifts westward aboard its tectonic plate at the steady rate of one centimeter a year, one meter a century, one kilometer every millennium (a nice distance in metrics), or six-tenths of a mile each thousand years. If this movement had been constant since earth's birth (which it wasn't for reasons that will become clear), and if the Smithsonian had been built immediately, the Museum would have circled the globe 117 times at that minuscule pace over that enormous reach of time.

Imagine time in a linear fashion. Were this book to devote a single page to each year of earth's life, its leaves would pile up more than 181 miles thick. Bound and placed on an impossible shelf, the anthology would use the Washington Monument here for one bookend and the George Washington Bridge outside New York City as the other. If it devoted one page to each proverbial human lifetime of three-score-years-and-ten since earth began, the tome could be stored within San Francisco's Golden Gate with scant room to spare. But of course California wasn't yet a gleam in the cosmic eye when earth began; nor was mankind or anything else biological. So far as anyone can discover thus far, there was no life at all for almost the first billion years. The oldest rocks that still survive were formed a billion years after earth began, time enough for a stack of pages 39 miles

WEST VIRGINIA
WEST VIRGINIA
WEST VIRGINIA
WEST VIRGINIA
WEST VIRGINIA
WEST VIRGINIA
Dunkard
Dunkard
Dunkard
Dunkard
Dunkard
Dunkard
Dunkard
Dunkard
Dunkard

tall in the unabridged east coast edition or an awesome 14,285,714 pages in the condensed western edition. Thus, if earth's dynamics proceed at an imperceptible pace from man's puny perspective, they have had an immense stretch of time to do their work.

If no rocks survive from the first fourth of earth's life, what petrified "pages" have geologists read to explain earlier events? Answer: the meteorites which have been colliding with earth since the beginning. (They have been collected since Joseph Henry took custody of Smithson's personal effects, which included several fragments.) Of course, no meteorite remains from earth's molten days, so the evolution of the early planet does remain a matter of careful scientific conjecture. Importantly, however, meteorites, were formed by the same event that made earth. They contain clues to the origin of the solar system and every body in it.

In contrast to the visible phenomenon of a meteor—a shooting star—meteorites by definition are the cosmic objects that fall to earth. They've had a checkered career in science. By the time of Smithson's birth some responsible collectors knowingly threw away celestial rocks hoping to end the crude superstitions of common folk who said these stones rained down from heaven. By the turn of the nineteenth century the consensus was changing, albeit slowly. A meteor shower was reported by the French Academy in 1803 though many members still doubted reports from peasants who'd collected pieces of the L'Aigle meteorite—doubted them because they were peasants. Four years later when two Yale men described a fall near Weston, Connecticut, no less an Enlightenment savant than Thomas Jefferson purportedly said he thought Yankee professors would lie before the sky would drop rocks.

But hard evidence, first reported by ninth-century Chinese, kept appearing in the Occident. Smithson's collection contained 14 fragments which were analyzed before a fire destroyed them and gutted three of the Castle towers in 1865. They had become objects of valid scientific interest. In 1882 George P. Merrill became Curator of Geology and "he thought everybody with a meteorite should give it to him," Curator Emeritus Edward P. Henderson told me. The whole Division of Mineralogy, part of the geology department, which then included paleontology, comprised

Opposite: *Not restricted to visually spectacular material, the geology collections contain hundreds of cases of seemingly ordinary rocks which reveal more about earth's history, structure and composition than gems do. These West Virginia sedimentary rocks happen to contain Devonian Period plant fossils.*

Glazed by frictional heat from falling through earth's atmosphere, a poundcake-sized fragment of the Allende meteorite bears a layer of dark fusion crust. The interior, exposed by a terrestrial fracture, contains light inclusions rich in calcium and aluminum. Radioactive dating sets Allende's age at 4.6 billion years, the same age as our solar system. But isotopes in the inclusions lead some experts to date them even earlier—remnants of material that existed before earth's birth.

Opposite: *Component minerals in a magnified thin section of an Allende meteorite fragment display a panoply of diagnostic colors under polarized light.*

only two people after Henderson arrived as the junior member in 1929. But it found an aggressive exponent in Stuart H. Perry, vice president of the Associated Press and the Museum's benefactor. Fascinated by fossils as a boy, as a newspaperman Perry displayed whatever natural curiosities readers brought in to his Michigan paper. A natural science buff who baldly used the press to popularize his interests, he collected meteorites in particular and traded samples with Merrill. Henderson, in turn, catalogued, described and shelved them. What they could reveal about earth's origins remained to be seen after more sophisticated analytical techniques replaced cruder chemical assays. "Measurements made 50 years ago were junk compared with today's," Henderson says.

Swapping surplus material, "borrowing" for indefinite terms, begging donations for purchases and collecting on their own, Henderson and his successors acquired samples of at least half the 2,500 known meteorites, including the Weston and the L'Aigle. (Most are named for the places where they fell.) The most successful single recovery effort occurred in 1969 after a spectacular shower occurred near Allende, Mexico. Roy S. Clarke, Jr., the curator, and Brian Mason rushed to the site, enlisted local residents in the search and gathered hundreds of pounds of fragments.

The significance of meteorites, emerging slowly over centuries, became clear with new theories about the genesis of the solar system—theories that were in turn clarified by these missiles or "free space probes" as Clarke calls them. In 1772 German astronomer Johann E. Bode popularized an arithmetic formula that accurately described the previously unexplained spacing of the planets' orbits. Progressively wider intervals separate the sun from Mercury, Mercury from Venus, Venus from Earth and Earth from Mars. Then a vacancy appears where a planet "ought" to be and the progressive spacing becomes regular again with Jupiter, Saturn, Uranus, Neptune and eventually Pluto. (Neptune was only discovered the year the Smithsonian was chartered; its orbit was analyzed in the second issue of the *Contributions to Knowledge* series.) As for the hiatus between Mars and Jupiter, by 1900 it was known to contain at least 500 asteroids which orbited the sun en masse without incident, so far as anyone could tell.

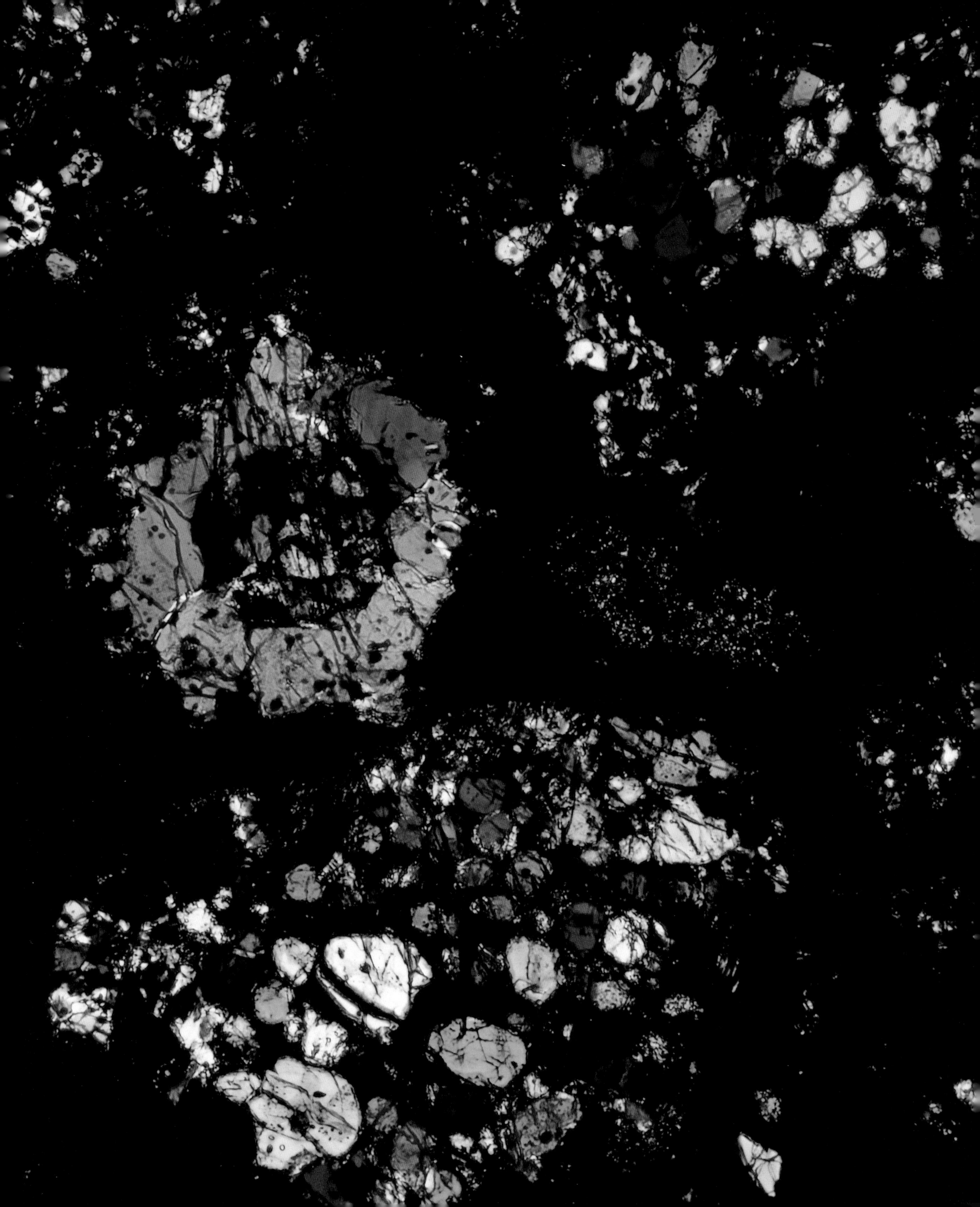

Now leap, if you will, to 1970 when a meteorite fell near Lost City, Oklahoma. Its descent had been photographed by the Astrophysical Observatory's five-year-old Smithsonian Prairie Network, an array of robot cameras scattered across seven states which scanned the entire sky over the Great Plains every night. The films were processed and the meteorite's touchdown point plotted so accurately that the technician who went searching found it on a snow-covered road 450 miles from the nearest camera and 750 yards from the predicted spot. The 21-pound meteorite was dispatched to the Museum and its trajectory traced back to its solar orbit from the carefully calibrated films. This trajectory and others confirmed that the asteroid belt probably spawns all meteorites that reach earth, after some local event jostles them out of orbit. Radioactive decay data, a dating technique employing the known rates at which unstable forms of elements change to stable ones, indicate that all meteorites are generally the same age, about 4.6 billion years old. Formed at the same time, they all spent the interim orbiting in the same vicinity, namely where Bode had calculated a single planet should be. "There wasn't enough material to make a decent planet" between Mars and Jupiter, Dr. Mason believes. What was there—from microscopic particles to planetesimals a few hundred miles in diameter—revolved where more of their kind might have become a planet.

The astrophysical consensus holds that our entire solar system derives from one event, possibly the collapse of a gas cloud or stellar explosion, a supernova. The laws of motion compelled most of the resulting debris to scatter through space while a fraction spun into a shallow disk like a nebulous Frisbee. Since all the condensing gas and particles revolved in one direction at similar speeds, they often collided gently and adhered to each other. As small bodies aggregated to become larger, their gravitational fields grew exponentially and they attracted new material from their orbital vicinities. In other words they snowballed until each young planet dominated its own orbital space around the epicenter, now the sun, which contains 99.9 percent of the material in the entire system. When each nascent planet became a critical size, its own gravity compressed its

The unusual Tucson Ring meteorite, measuring 3 by 4 feet and weighing 1,500 pounds, was known to Spaniards who colonized what became Arizona. Later used as an anvil, it was finally freighted overland to California in 1863 and by ship to the Isthmus of Panama, across Panama on land and then by ship to Washington and the Smithsonian. The high silicon content of its iron indicates formation in an environment with abnormally little oxygen—even by meteoritic standards.

Opposite: *While meteorites still tumble from space, most of the moon's surface scars date from the "heavy bombardment" period around 4 billion years ago, which must also have brought a veritable rain of rock to earth. Tectonic action and this planet's atmosphere erase such scars over time, but on the weatherless moon impact craters remain almost perfectly preserved. They were often mistaken for volcano cones from the time of the first telescopes until their origin was proven for good by the Apollo space missions, one of which took this picture of the moon's far side.*

material tighter. Energy released by accretion, and the decay of short-lived radioactive elements, provided heat enough to melt. Once accretion and radioactive decay rates diminished, the heat dissipated and surfaces cooled to form hard crusts. Thus the four "terrestrial" planets—Mercury, Venus, Earth and Mars—were born along with some of their satellites, notably our moon. Then came the relatively impoverished asteroid belt in a region that Dr. Mason believes might be a kind of boundary in terms of water's behavior. Then came the series of huge distant bodies composed largely of frozen gas.

Analysis of all meteorites in the Museum's collection helps confirm the single genesis of all planetary and asteroid material. Given different categories of meteorites, all are composed of the same materials found on earth and its sibling bodies. In the aggregate the common stony chondrites are composed of the same proportions of elements as the earth itself. Indeed, Mason says these meteorites have revealed more about earth's total composition than terrestrial studies, since only the crust is accessible and the core and mantle cannot yet be studied directly.

Similarities aside, the elements that make up meteorites have been combined into many different minerals, some of them unique, because they encountered different conditions such as original melting temperatures, pressures and cooling rates. The Allende also contains white and gray inclusions, particles ranging in size from peppercorns to walnuts, that might be older than the rest of the material. If further study, or an as yet uninvented tool, confirms this, it follows that they are remnants of some star elsewhere that antedates the solar system.

The possibility of confirming matter older than the solar system is just one of the challenges that has revitalized geology since space exploration began. Some of the impetus came from NASA, which financed new kinds of terrestrial and celestial studies in order to achieve President Kennedy's 1961 goal of sending a man to the moon within the decade. Some pieces of the puzzle appeared as palpable results of various space projects and missions. Other pieces appeared serendipitously, like Antarctica's lode of meteorites, which Japanese scientists found while embarked on other work. (The polar continent hasn't been struck by

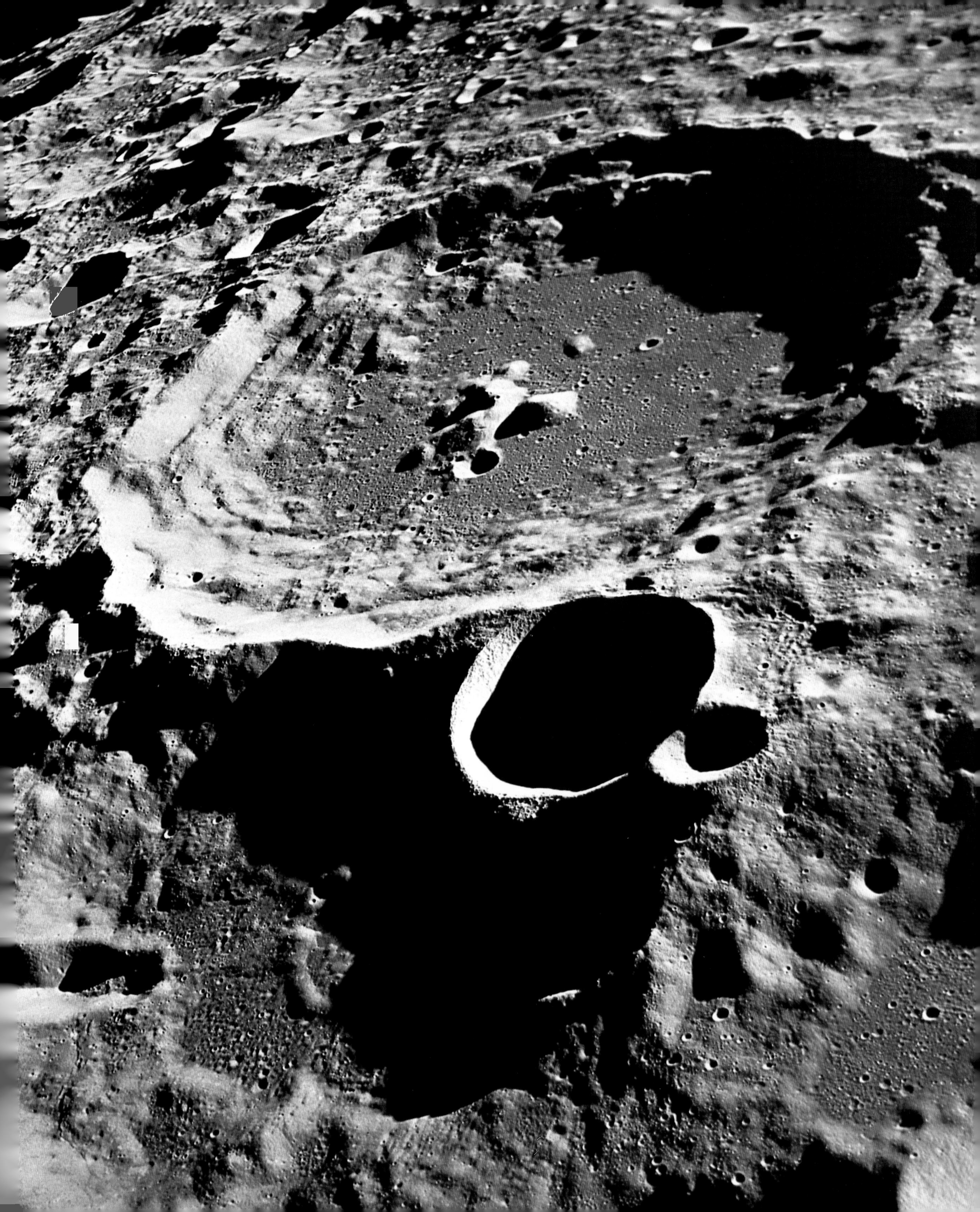

Brought to earth by Apollo XV astronauts, this bit of "young" basalt seeped up to the moon's surface and solidified some 3.3 billion years ago. The sample illuminates lunar history because it has not been oxidized by terrestrial air, which could alter its chemistry. NASA mounted and sealed the fragment in a display case filled with inert gas to keep it molecularly pristine.

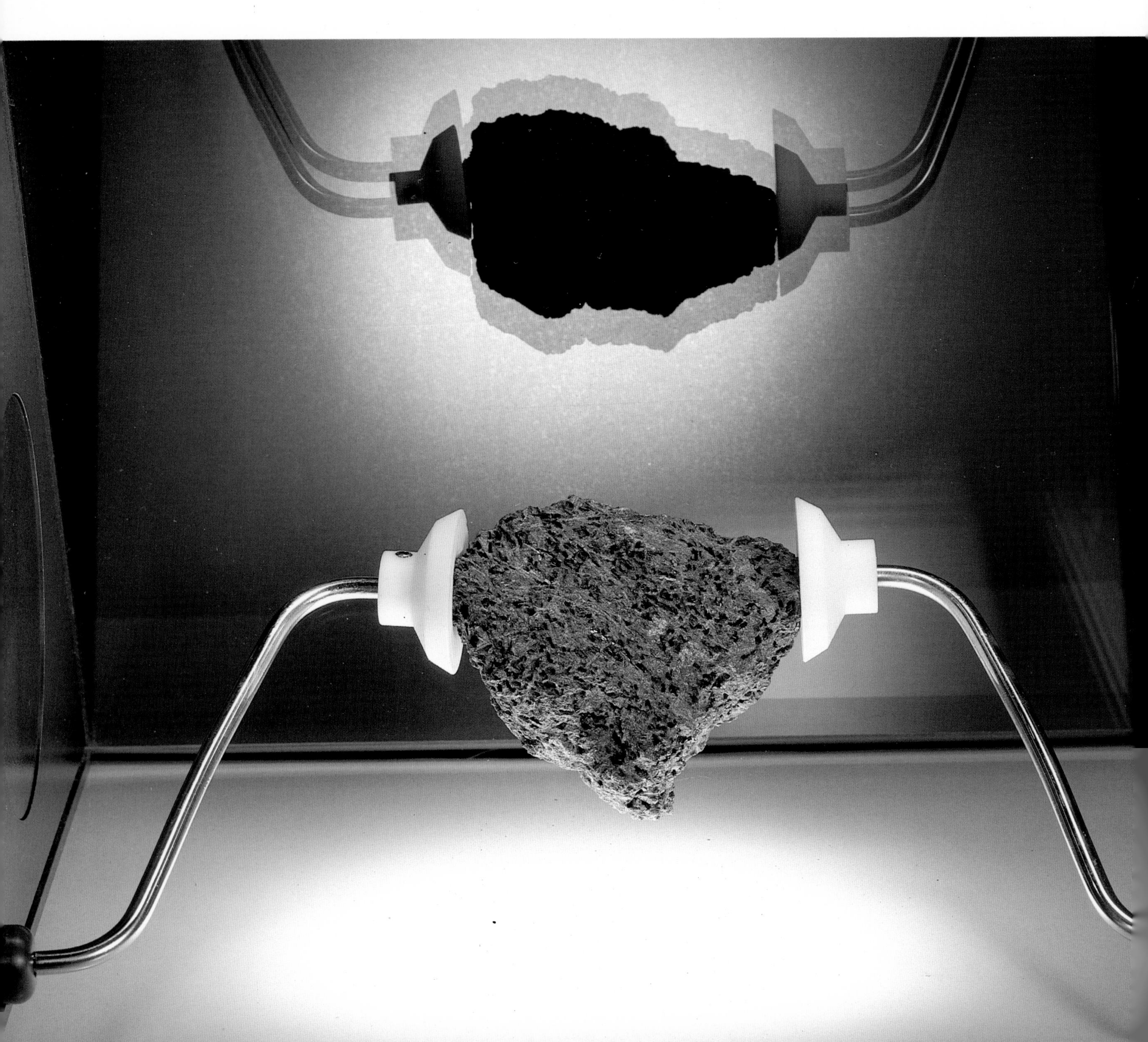

meteorites more often than the rest of the planet. Its ice just preserves them better, while glacial movement and shrinkage tend to collect the crop of many millennia and concentrate it.) Still other confirmed concepts marked advances of science, as discoveries in one field contribute to new understanding in another.

Continuing earth and planetary work confirmed that while known meteorites remain relatively rare, far more of them regularly strike the earth without being found. Perhaps 500 of these missiles (as big as baseballs or bigger) strike the planet's 197,000,000 square miles a year. Given those numbers and the unspeakably long odds they represent, it's almost surprising that as many as two have landed very near people in recorded time. One of those fell on July 4, 1917, behind a farmer's barn in Colby, Wisconsin. Moments later the man found it buried a couple of feet in the ground and—contrary to popular expectation—said it was covered with hoarfrost, an observation that confirmed his report. (Most people expect a meteorite to remain hot long after its fall. In fact only the surface heats up as it falls, like the Apollo vehicles, which employed the same mechanics to keep astronauts from burning up. This surface heat quickly dissipates. A meteorite's interior temperature remains very low from the deep freeze of space.) The second notable witness, Mrs. Hewlett Hodges, was hit on the hip after a meteorite crashed through her roof while she napped after lunch in Sylacuaga, Alabama, in 1954. That event, which left the lady looking bruised and bewildered as seen in a photograph, remains the only confirmed human hit on record.

Chicken Little's fears notwithstanding, two related factors prompted geologists to reconsider the frequency of meteorite falls. One was a debate joined in Galileo's day when people first got a magnified look at the moon. Some Renaissance men and subsequent scientists alike believed the pockmarked surface showed cone-shaped volcanoes. Others argued that meteorites had done the damage. If the latter case were true—as closer observation would prove—how could seemingly unscarred earth, a larger passive target, have escaped being hit more often? The second related factor—a neglected one—was time.

This year's odds against any single acre being struck are better than

Overleaf: *A thin section of the same moon rock, illuminated by polarized light, displays a variety of colors under the microscope.*

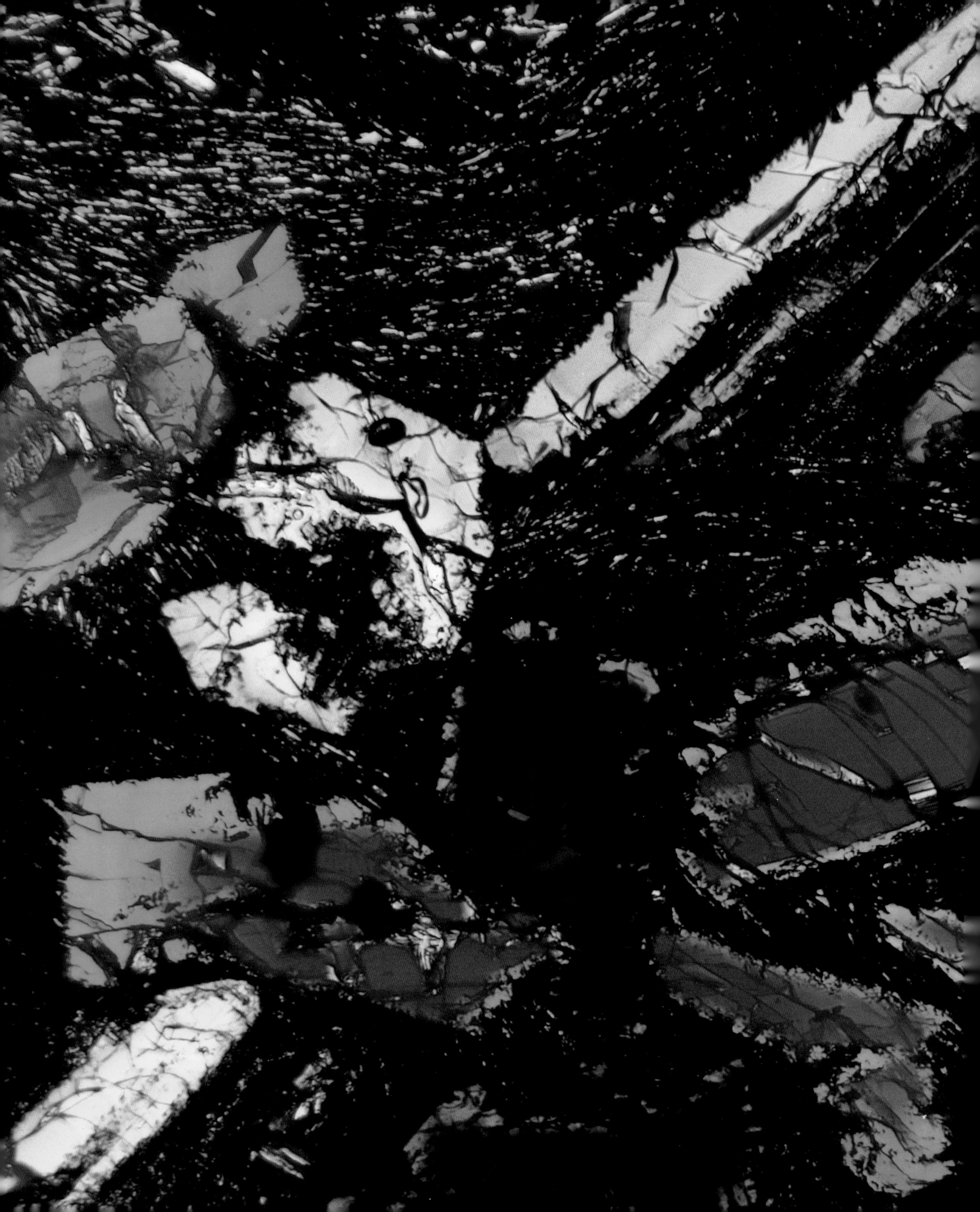

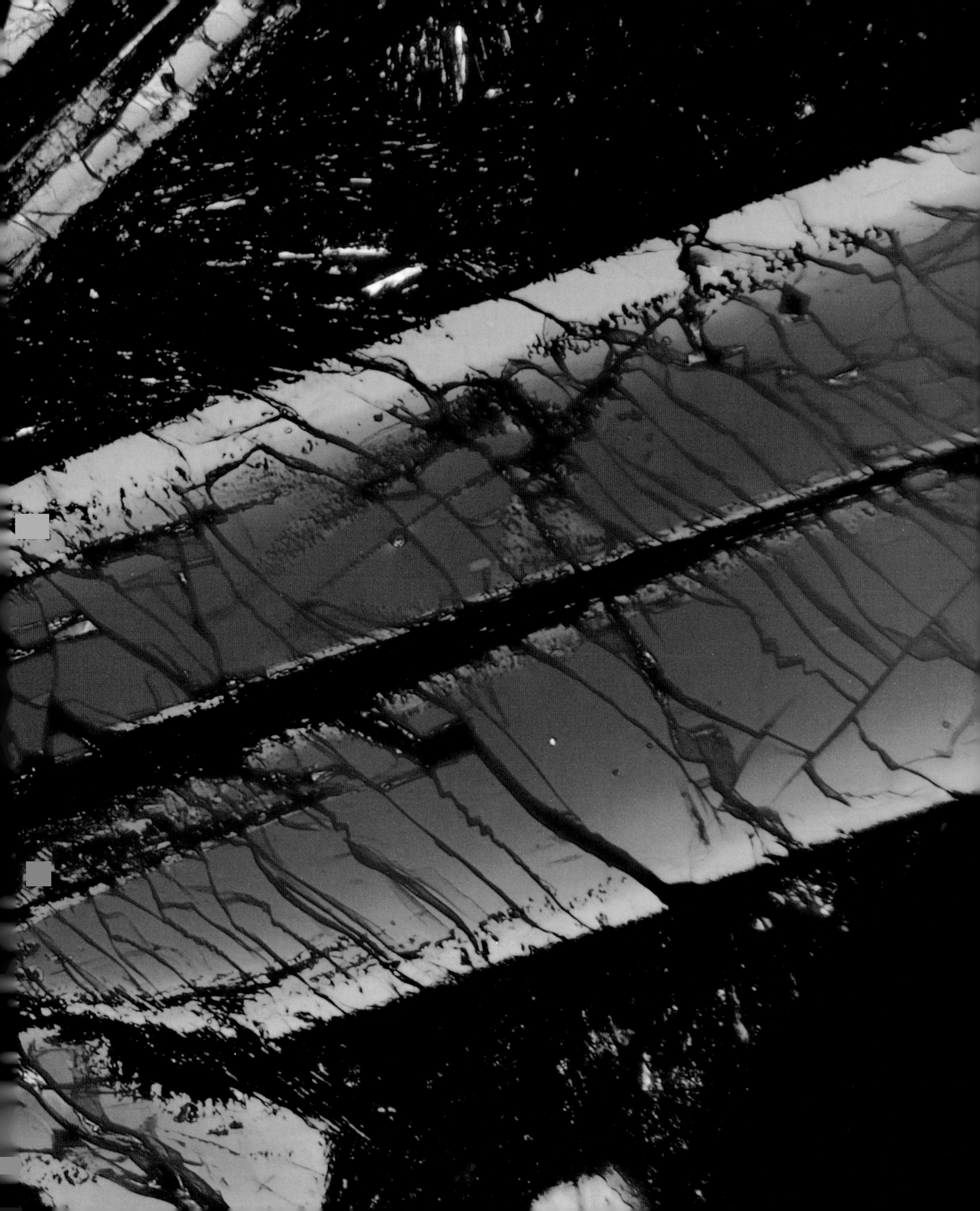

While electronic technology offers very precise assay techniques for minute rock samples, Museum scientists still employ "wet lab" chemistry for bulk analysis and other tasks.

125,000,000 to one. But said another way, every acre of earth should have been struck by at least a fist-sized meteorite every 125,000,000 years, or 37 times since earth's beginning. The Apollo missions confirmed beyond any doubt that meteorites caused the craters on the moon, some of them more than 200 miles across. The full moon's lighter features in the highland regions are meteorite impact scars and debris from a "heavy bombardment period" between 4.1 and 3.9 billion years ago. The lunar lowlands, or the Man in the Moon's dark cheeks, are lava plains. In these layers lies evidence of vulcanism, but not the explosive sort which some scientists hypothesized from the mountainous cones. Rather, these smooth regions represent impoundments of basalts which seeped up as lava from the moon's interior through fractures—some of them undoubtedly caused by meteorite strikes after the crust first formed. The *maria*, Latin for seas, turn out to be aptly named in a figurative sense. Once seas of lava, they crystallized into rock. Radioactive dating of material from different parts of the maria show that the older basalts came from lava produced at lower temperature than the younger rocks. Thus, as time progressed, younger volcanic flows came from deeper within the moon, suggesting that it has cooled and its exterior layers solidified considerably since its origin.

Dating also shows the oldest lunar rocks to be about half a billion years older than any terrestrial rocks yet dated. Preserved by the near vacuum of space, lunar rocks remain in a nearly perfect state of preservation. It turns out that the lifeless and rigid moon—lacking both earth's erosive atmosphere and dynamic crust—constitutes the ideal conservator of relics from the early period of intense bombardment and from later meteoric events.

Seismic surveys of the moon suggest concentric layering: possibly a small metallic core surrounded by a thick layer of partially molten glassy silicates surrounded by a thick, solid mantle. In these details the moon's structure differs from earth's, possibly because the smaller body cooled faster and more completely. But the moon's crust and earth's contain the same materials in similar proportions. Unidentical twins, they must share other characteristics and experiences.

Extrapolating from the number and size of the lunar craters suggested surprising things about earth's past. On a statistical basis, earth could have been hit by enough meteorites to have 500,000 craters larger than three-fifths of a mile in diameter, more than 10,000 craters larger than 6 miles in diameter and more than 200 craters larger than 60 miles in diameter, plus a few real giants. If this were so, where is the evidence? Much of it has been erased by oxidation, wind, rain and erosion, functions of the earth's atmosphere (the mix of gases which escaped the weaker gravity of the smaller moon). In a few places clues remain to the battering earth has received: Canada's Sudbury Basin and South Africa's Vredefort Uplift, which both measure more than 80 miles across; Arizona's mile-wide Meteor Crater, which is only 25,000 years old. Created by an object that weighed between 18,000 and 144,000 metric tons, it was formed by the force of a 4-megaton nuclear bomb. (This is not to suggest that a thermonuclear explosion occurred, but that an equivalent amount of energy was instantaneously released when the meteorite landed and shattered.) The vague weight estimate derives from the uncertainty about this object's speed. Meteorites enter earth's atmosphere at velocities of 10 to 20 kilometers per second, ten to twenty times "faster than a speeding bullet"

Opposite: *Broken open like a coconut, a geode reveals its interior. Light shining through the specimen's impure surface imparts rich color to the white quartz.*

in the old *Superman* vernacular, or 22,000 to 44,000 miles an hour. And the bigger one is, the less earth's atmosphere slows it down. Anything over about 1,000 tons doesn't decelerate at all before touchdown.

The question remained, if events like these were relatively common over time, what effects would they have had on earth during its development? During the heavy bombardment period the larger meteorites might have contributed to the depression of ocean basins and elevation of highland masses, perhaps primitive continents.

In some respects the search for ancient evidence resembled a pulp detective's hunt for the *corpus delicti* after learning that a crime, in fact, had occurred. Once geologists decided that many impacts must have occurred after earth developed a stable crust, they discovered a new class of metamorphic rock, the remains of meteorite strikes. Normal metamorphism requires geologically modest pressure and temperatures up to about 1,000° C over some millions of years. But a meteorite impact raises rock temperatures to 10,000° C and creates instantaneous pressure several thousand times higher than normal. Shock waves radiate long distances through the terrestrial rock at rates of a few miles a second. Both heat and pressure waves irreversibly change the target rocks. The impact melts many tons of material and throws some of it into the air, perhaps with enough force to leave the atmosphere before falling back.

Impact metamorphism explained whole new families of rock. It also suggested why microscopic diamonds appeared in some fragments around Meteor Crater when most terrestrial diamonds form 100 miles beneath the surface, where temperatures reach 3200°F and pressure builds to 2 million pounds per square inch. Recalling Hutton's old rule, it might seem that meteorites making diamonds contradicts the Theory of Uniformitarianism, or that giant meteorites themselves might prove a Theory of Catastrophism. These falls would certainly be catastrophic for anyone in the vicinity, but the fact remains that catastrophes are part of the natural scenario. While impossible things don't happen, events with astronomical odds must occur within an astronomical period of time. As Brian Mason expresses the rule with a scientist's blend of precision and understatement, "If an event is very improbable, it's bound to happen in three or four billion years."

This same probability factor explains much about mineralogy from meteorite diamonds to the magnificent variety of minerals. More than 3,500 species occur on earth and the collections contain nearly all of them—given that nearly a hundred new ones, most of them microscopic, are identified annually. The reason for this variety lies in the nature of a mineral, by definition a "naturally occurring, homogeneous inorganic substance having a definite chemical composition and characteristic crystalline structure." But here's the rub: a finite number of chemical components can combine in a great many ways. Many species of distinct minerals have identical chemistry but different crystal structure, diamonds and graphite for example. Others have identical crystal structure (there are only six or seven geometric crystal shapes—depending on who's counting) but different chemistry. Calcite, a common crystalline form of calcium carbonate, is the central component of limestone, marble and chalk. Replace its calcium with zinc and you've found a piece of zinc carbonate, i.e., smithsonite. Impurities in corundum (aluminum oxide) or minor substitutions of metals make the difference between a ruby and a sapphire, which is simply a ruby of any color except ruby red. Beryl, called an emerald when it's emerald green, comes in every conceivable color. Many of them are named, many more anonymous. The chemical formula for the tourmaline group suggests how many distinct tourmaline species can occur with the addition or substitution of a single substance: $(Ca,K,Na)(Al,Fe,Mg,Li,Mn)_3(Al,Cr,Fe,V)_6(Be_3)Si_6O_{18}(OH,F)_4$.

If the 10,000 gems represent one end of the collection's spectrum, the 180,000 specimens of plainer rocks, which William Melson oversees, represent the other. One of the senior scientists who began his career as a boy haunting exhibits and pestering curators with questions, Dr. Melson later commanded research activities aboard the *Glomar Explorer* on two important voyages. He succeeded in drilling closer to the layer below earth's crust, the mantle, than anyone before when he bored 1903 feet through the ocean floor off the Azores. The work took place at sea because the crust is thinner (and younger) beneath oceans than on the continents.

Melson's special expertise involves the manifestations of plate tectonics, the dynamics of earth's surface which notably appear as volcanoes,

These pyrite and marcasite concretions formed after mineral-bearing water passed through voids in existing stone.

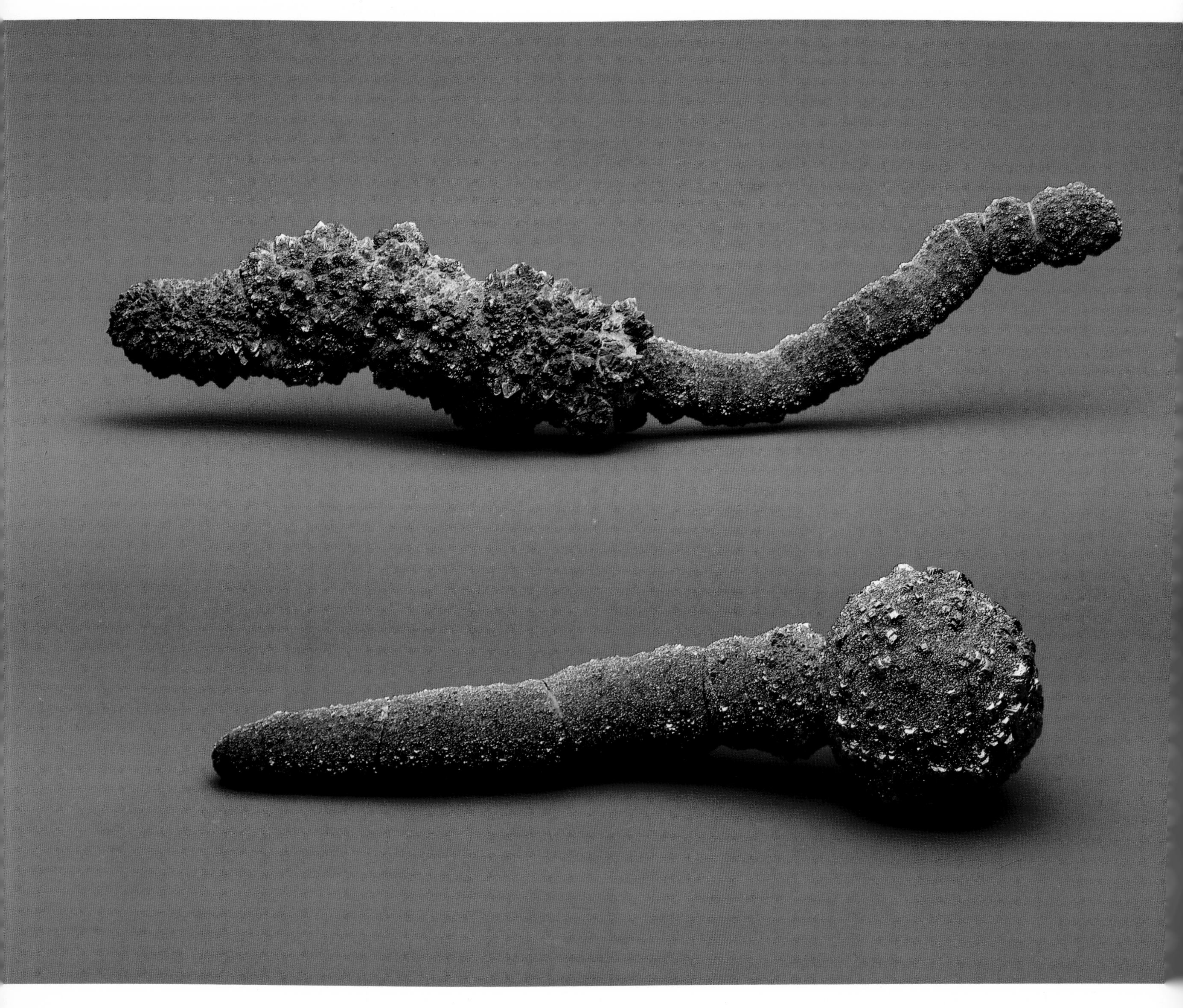

Left: *Agate, one variety of silicon dioxide, forms in geothermal regions. This complex specimen, found in a Mexican mine, filled a void in the surrounding rock as water containing dissolved silica passed through. The solution left layer upon layer of crystal material on the walls of the shrinking space until it closed.*

Below: *The large uncut emerald, mineralogically speaking a green beryl, is composed of beryllium, aluminum and silica. The characteristic color of this specimen derives from minute amounts of chromium—on the order of a few parts per million—locked within the crystal structure. Beryl is found in other colors as well. When the impurity is manganese, the crystal is usually red; when iron the crystal may be blue or yellow.*

Overleaf: *This mineral pom-pom, a scolecite crystal cluster from India, is made of calcium, aluminum, silica and water.*

This amber-colored arrangement of calcite crystals, about the size of a platter, recently came from a natural Tennessee cave within a lead and zinc mine. The calcite, a form of calcium carbonate, apparently precipitated out of the surrounding limestone, which has the same chemical composition but a different form.

earthquakes and the inexorable drift of continents. These phenomena derive from the little-known mantle, a 1,800-mile-thick mix of heavy silicate minerals which surrounds the liquid iron outer core and the solid iron inner core. Like Silly-Putty, the mantle has odd physical properties, behaving like a flexible plastic under sustained pressure in its upper part, but like a brittle solid under sudden shocks. The tectonic plates, which compose earth's surface, crust and uppermost mantle, literally float on the deeper mantle and move about irregularly. Scientific acceptance of this fact occurred about as fast as continental drift. The notion, first proposed in 1915 to explain the "fit" of continents, was widely denied for nearly 50 years because no one could explain the driving cause. Now a consensus holds that convection-like motion in the mantle makes the plates move. If the tectonic plates move and collide, it follows that there must be places where the plates are separating. Observations confirming this fact provided the keystone for convincing tectonic theory.

The Atlantic Ridge looks like a submarine mountain chain running

irregularly from the latitude of Cape Horn to the Arctic Circle. It is a "seafloor spreading zone," a region where plates are drifting apart. As they slowly separate molten magma seeps up from the underlying mantle to fill the gap. As magma rises, it cools and solidifies and its newfound crystal structure records the magnetic polarity of earth. That polarity has reversed itself many times over the eons. Sensitive recorders aboard a research ship measured the width of each magnetic stripe, namely the breadth of seafloor which crystallized during a period when north was north, then north was south, and so on. The pattern of magnetic stripes proved to be symmetrical; each stripe on the east side of the ridge had a mirror image on the west side. The only explanation is that the plates have been slowly spreading apart. As this occurs the magma wells up between them and solidifies, becoming part of each plate's trailing edge.

As a plate continues to drift, its leading edge encounters the next plate and may "subduct" (be pushed downwards) into the mantle where the resulting friction creates melting heat. In this kind of vicinity, volca-

Overleaf, left: *Gold also emerges within material from the mantle and when this occurs it too may crystallize into growths like these within quartz. The gold and quartz have segregated themselves because gold has an affinity for itself while oxygen and silicon have an affinity for each other. When the latter two combine into silica and crystallize as quartz, the emerging crystal cannot accommodate gold. When erosion, landslide, earthquake or another event disturbs the crystalline lode, the gold may be freed from its rock matrix and washed into streams. Extremely malleable, it tends to remain intact albeit rolled and pounded into nuggets of almost any size. Witness the five-pounder on page 130.*

Preceding page: *This wulfenite, a mineral composed of lead, molybdenum and oxygen, comes from the San Francisco silver mine in Sonora, Mexico.*

Below: *The mineral Vanadinite from*

Mibladen, Morocco.

Opposite: *This gypsum flower evidently grew out of its base into a void in the surrounding stone. Gypsum combines calcium sulfate and water.*

Opposite: *A geode's remarkable history begins deep within the earth whence molten magma rises as volcanic lava, which crystallizes as basalt. When the lava begins to cool, it releases gases which coalesce as bubbles. Flowing more slowly as it solidifies, the lava stretches and flattens any captive bubbles which remain as voids after the rock sets. Water seeping through the still-hot stone becomes superheated to dissolve silica, among other things, and make the stone more porous. With time, further cooling, and migration of water through the hardened bubble, dissolved minerals crystallize on its inner walls. Silica formed the shell of this rare star geode, first as a common variety of white quartz, then as the rarer amethyst with the arrival of a new impurity or changing temperature. Later calcium carbonate crystallized to form three bursts of calcite, which in turn became covered with more amethyst as conditions or dissolved components changed yet again.*

noes are active and major earthquakes occur. Where neither plate subducts, mountains rise. Where two plates slide past each other, earthquakes occur but volcanoes are rare. In some active regions all three phenomena occur with catastrophic consequences for anything living in the neighborhood. Other areas seem simpler.

The Himalayas, for example, are rising today. They mark a collision zone between the huge Eurasian Plate, which slides slowly southeast, and the Indo-Australian Plate, which is moving northeast. True to James Hutton's theory, this kind of action has built high mountains before; for example, the Appalachians, which stretch from Alabama to Canada. Now shadows of their former selves, these mountains are thought to have been eroding for the past 200 million years. For a longer time before their rise, they were part of a primordial seafloor.

The Andes also rise as the South American Plate and the largely submarine Nazca Plate meet nearly head-on. Here the oceanic plate subducts beneath the thicker and lighter continental one along four-fifths of South America's west coast. Where it meets the mantle, the friction may help to create a distinct melting zone of temperatures reaching 2200° F (1200° C). Growing reservoirs of molten magma press up through the Andes and escape periodically through volcanoes, whose eruptions are powered mainly by superheated water vapor.

Further north, just off the Pacific Northwest coast, lies the Gorda Plate. One of the narrowest, it reaches barely 300 miles from the spreading zone of a Pacific ridge. As it encounters North America, it too subducts to a melting zone. Its material contributes to the lava which recently exploded from Mount St. Helens, just one of the few-million-year-old Cascade Range's historically active volcanoes. During field studies before and after the eruptions, Melson learned that this lava contains unusually high amounts of water. This may be because the narrow undersea plate subducts unusually fast and at a steep angle. In any event, Mount St. Helens has exploded many times before and will again, though quite possibly another of the Cascade cones will beat it to the punch. After each eruption releases subterranean pressure, lava cools and seals the vent, until the pressure rises enough to blow off the lid again.

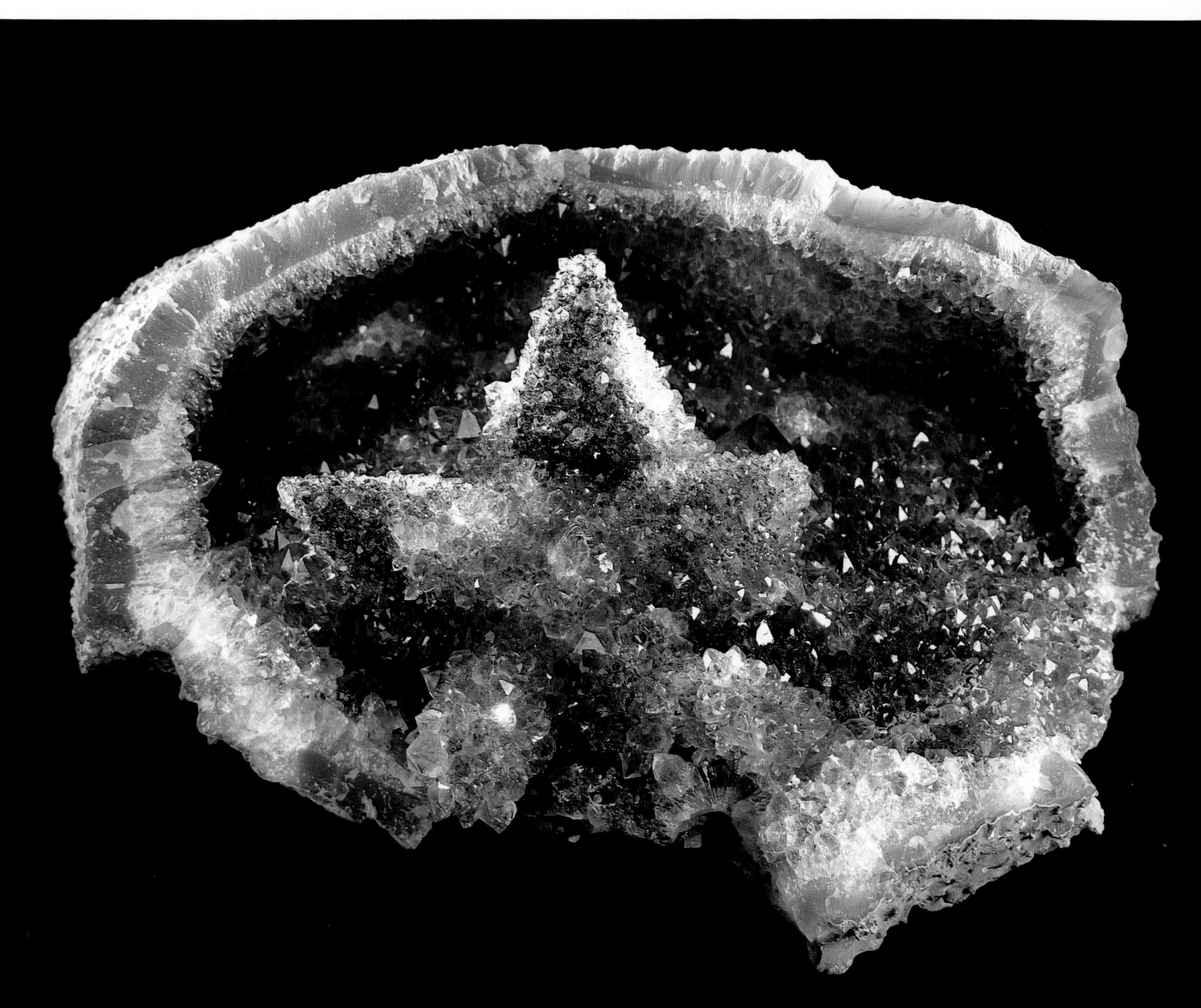

Opal is not a true solid but technically a fluid, like glass, combining silica and water. Sometimes unstable, the iridescent gem material will crack if the water escapes. In its natural state, opal precipitates out of groundwater into rock crevices from which it may then be cut.

Right : *Romanian stibnite crystals (antimony sulfide) grow like clumps of steely crabgrass dotted with the dewy barium sulfate barite.*

Following pages: *Manganese carbonate grows as spectacular scalenohedral crystals featuring triangular faces with unequal sides. This rhodochrosite came from a mine in South Africa's Kalahari Desert.*

Most submarine and terrestrial volcanoes appear near the edges of tectonic plates but a few occur over unexplained "hot spots" in the lower regions of the mantle. Their causes remain obscure but their results are as apparent as the Hawaiian Islands. A hot spot lies dormant for a time; then magma rises up through the crust of the ocean floor; then the hot spot goes into a kind of remission while the Pacific Plate continues its slow drift. This results in a series of volcanoes which rise from the sea floor and trace the movement of the plate over the hot spot. When one stops growing before breaking the ocean surface, it forms one kind of seamount. When it continues, it's called Maui or Oahu. Over the eons the Pacific Plate has rotated nearly 90 degrees as it drifted north and west, according to a Museum ichthyologist who studies the evolution of fishes in terms of continental drift. (One result of seafloor spreading: a single marine habitat can be split and its parts separated by greater and greater distance. Over time a single shallow-water species may be divided into two populations, then become two new species.)

When two plates slide past each other, the inevitable result is earthquake. Witness California's notorious San Andreas Fault. Just offshore, the Pacific Plate slowly inches northwest while a slight rotation of western North America moves most of California southeastward. The fault itself marks the division between the two plates. As each moves, enormous forces build up in the abutting rock surfaces which reach deep down through the crust; each edge bends backwards like the tip of a schoolboy's ruler about to launch a spitball. The results can be obvious, as when a straight road or fenceline crossing the fault starts making an abrupt and progressively sharper S-curve. When enough stress builds up to break the two faces' grip on each other, the two plate edges spring forward past each other. The relative movement may measure tens of feet; the road or fence shears off over the fault and the two segments end up that far apart, each of them a straight line again. The energy released by such an event opens deep narrow chasms, shakes the local region with unspeakable violence and sends shock waves around the world. In human terms the havoc is cataclysmic, but as James Hutton theorized, earth has sustained as much before and will again.

Opposite: *Though it might look like quartz, a very hard mineral, this is soft gypsum, a mineral composed of hydrated calcium sulfate.*

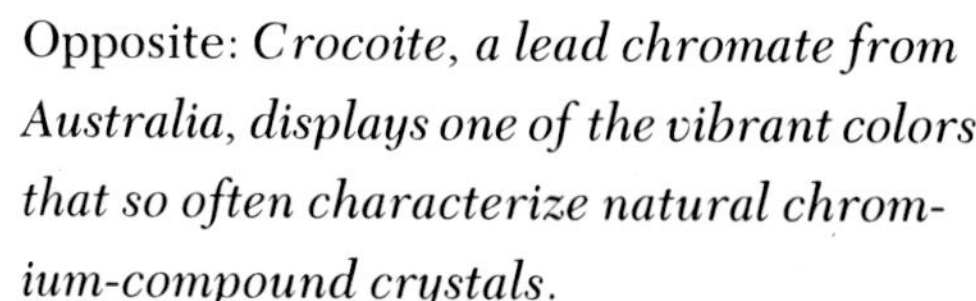

Opposite: *Crocoite, a lead chromate from Australia, displays one of the vibrant colors that so often characterize natural chromium-compound crystals.*

Left: *When a crystal exhausts one chemical component, another may replace it to start a different crystal growing piggyback on the original. Spanish miners unearthed this calcite-fluorite oddity. Both minerals combine calcium with another substance; calcite is calcium carbonate, fluorite a calcium fluoride.*

Bottom: *This opal is irregular in shape because the lapidary preserved as much as possible of the gem when cutting it from its natural rock matrix.*

A MUSEUM OF LIFE

If geology reveals the history of earth, paleontology describes the history of life here. The two sciences, focusing on the literally mundane and the organic, practice a kind of interdisciplinary pas de deux which is spectacularly displayed on the main floor in this Museum. Why this pairing? Because living things require a habitat and inhabitants either move, adjust or perish when the habitat changes—as all earth has physically changed. Thus the protean world, a place of inconstant climate, drifting continents, oscillating magnetism, erratic sea level, rising mountains, eroding land, corrosive air, etc., constantly dictates biological change. Less obvious, perhaps, life has had revolutionary effects on earth itself and its enveloping atmosphere. Today's air, for one example, owes its high oxygen content to photosynthesis, the process that green plants employ to convert solar energy into biochemical food. This dominating, noxious, highly reactive gas (hardly the life-sustaining element of choice on a theoretically ideal planet) is the byproduct of plant metabolism. When pioneer organisms began expelling it, the free oxygen polluted an atmosphere then probably composed of hydrogen, methane and ammonia. Thus greenery eventually remade earth's etherial envelope, which in turn presented new conditions to everything it touched: naked rock, exposed deposits of metals like raw iron, the dissolvent seas and the plants themselves. Second example: marine animals, which remove calcium and carbon from seawater for their own metabolic ends, have ended up creating considerable rock, especially limestone. For upward of 600 million years crustaceans, mollusks and other invertebrates have absorbed the chemicals, used them to form their shells, absorbed more as they grew, multiplied and died. Their remains, chemically transformed, became this common stone containing traces of its organic origin. While many geological processes occur independently of biological causes—continental drift, for instance—ancient life has dramatically affected the physical world, as the physical realm has framed life.

A rock layer, and the relic life it contains, can reveal much about each other. The stony matrix holds the organisms that lived together and offers clues to atmospheric or marine conditions; individual organisms, known from other deposits to have lived at specific times, can date the rock.

Trilobites, like Olenellus vermontanus, *which grew to 10 inches, ruled the world 550 million years ago. These marine animals numbered close to 10,000 species, possessed a pair of legs for each of their many body segments and lived on minute organisms and debris. But it appears that some crustacean upstarts arose during their heyday in the lower Cambrian Period and did the same things better. The trilobites died out, leaving no more survivors than the dinosaurs would nearly half a billion years later. The science of paleontology reveals evolution, radiation and specification as facts of life throughout biological history. So too has extinction been the fact of collective death for many times more organisms than inhabit earth now.*

Overleaf: *Free-swimming crinoids swarmed in the shallow seas that covered present-day Kansas 85 million years ago during the Cretaceous Period—so called for the chalk sediments that became fossil-rich slabs like this—of the Mesozoic (middle life) Era. Like modern starfish and sand dollars,* Uintacrinos socialis *possessed a five-part radial symmetry. The body was covered with calcium carbonate shell plates and propelled by waving the feathery arms. A few heirs—the sea lilies—remain, in ocean depths and around some coral reefs.*

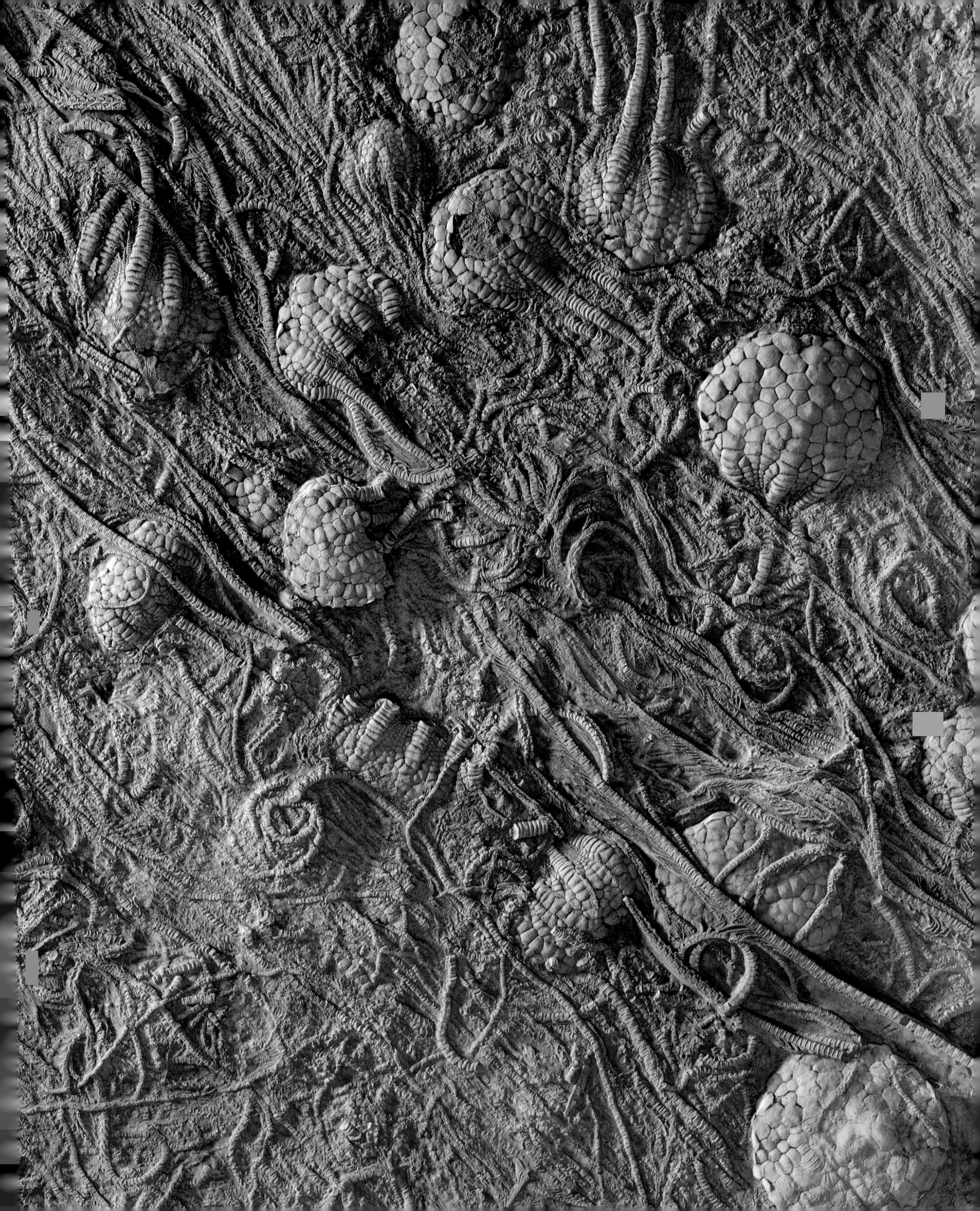

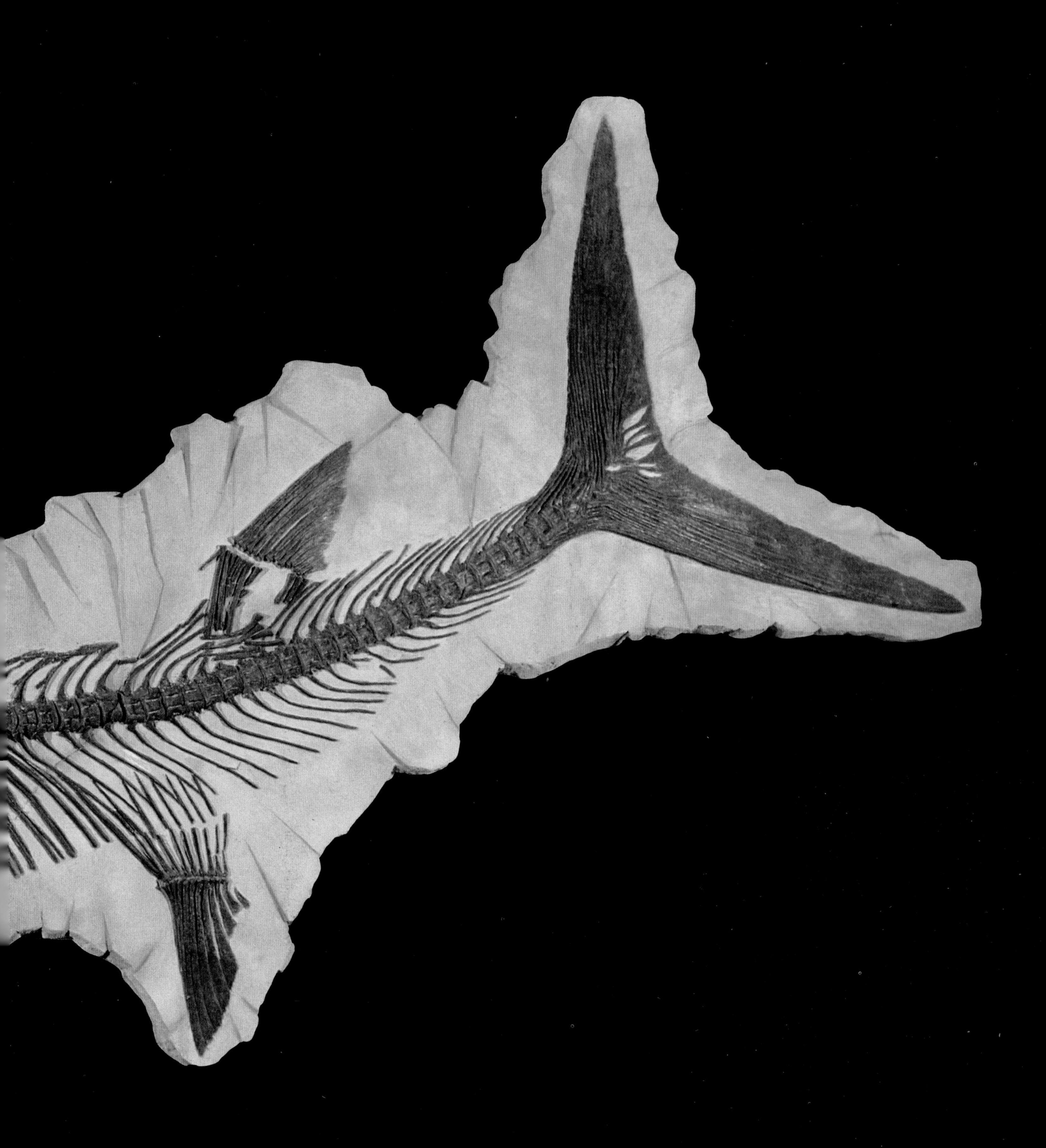

Preceding pages: *Chalk laid down more than 65 million years ago preserved this remarkable double fossil, a fish within a fish. The larger* Xiphactinus *species swallowed the smaller* Ananogmius, *which lies flexed within its rib cage. The food fish's nearly perfect preservation suggests it had so little time to decompose in the diner's belly that its very ingestion may have killed the killer. Then both were buried in the fine sediment of a shallow sea covering present-day Texas before scavengers could disturb the nested corpses.*

Opposite: *Like a witness in a murder trial, the fossil record can speak eloquently through its omissions. Fossil-bearing rock in North America contains specimens like this 50-million-year-old relative of modern puffbirds, a sluggish group akin to woodpeckers now limited to the tropics. During the Eocene, proto-puffbirds and related birds evidently occupied the niches now filled by passerines—crows, sparrows, flycatchers, warblers and such—the order comprising a large number of living birds. The multitudinous perching birds apparently didn't reach North America until 20 million years ago to start pushing the primitive avifauna aside.*

Through generations of scientific development, geology and paleontology tangled together inextricably in a single discipline until specialization gave their practitioners distinguishable identities. (In this respect these sciences resemble two biological species which evolved from a more primitive ancestor to fill neighboring niches in the habitat of science.) It was not until the Museum's reorganization of 1948 that the Paleobiology Department split off from the old Geology Division. Considerable overlap still occurs and always will.

The basis for the connection between the two sciences lies in a single specific: the history of life on earth is written in stone, and not limestone alone. Otherwise it could not have been transcribed at all. The fact remains that the narrative appears hard to read and annoyingly incomplete. Be patient. Science has been systematically trying to translate the billions-of-years-old fossil record for only a century and a half. One miracle lies in the amount we've learned; worldwide digging and patient deduction have produced intriguing details from mute evidence. Had the episodes been translated with greater artistry, this story would be the greatest epic in all literature. Its inherent plots, imagery, characters, self-caused tragedy, comic relief, intriguing asides, false leads and central wisdom are hardly matched in the combined works of Homer, Shakespeare, Dickens and all the anonymous authors of humanity's immortal myths.

If the medium is rock, the calligraphy is fossil henscratch. By definition fossils are remnants or traces of past life: the black-on-black imprints of the Burgess Shale Collection, the delicate brown ghosts of fishes and insects frozen in the tan matrix of the Green River material, the slabs of Cretaceous crinoids in the "Splendors of Nature" exhibit, the buried bundles of bones reconstructed as *Diplodocus*' remarkable skeleton. Despite the apparently enormous numbers of such relics seen in this Museum and elsewhere, fossils are rare. The odds favoring any organism's preservation are astronomical; we're lucky to have even representatives of many extinct groups. Most plants, animals and their remains are consumed by their contemporaries as the cat eats the mouse, the termite consumes the log, the crow disposes carrion, the mushroom transforms dung into delicacies. The nature of every discrete ecosystem we know

Opposite: *When a fine-grained sedimentary rock is split, it may reveal mirror images —called part and counterpart—of a single fossil. Source of the puffbird, the celebrated Green River Shale formation of the American Southwest preserved this herring, a species of* Knightia.

involves the conversion of a few inorganic substances into a series of innumerable organisms and the eventual return of their constituent parts back into the elementary pool. As the *Book of Common Prayer* so accurately describes the cycle: "Earth to earth, ashes to ashes, dust to dust," perhaps after a vital hiatus as a diatom, dinosaur, daisy or dentist. (In physical terms it is not quite beyond the realm of possibility to imagine this megacycle: A marine volcano like Iceland's Surtsey breaks the ocean surface and cools; life begins on barren slopes as airborne seeds take root, spiders parachute down and gulls alight; plants create new organic material from the young igneous rock and in turn feed insects and nearshore crabs, which in turn nourish birds and bottom-dwelling fishes, whose remains fall to the ocean bottom whence seafloor transport carries it to a subduction zone. Some of the molecules that flew up in the first volcanic burst—transmogrified a thousand times over a few millennia as rock, leaf, arthropod, vertebrate, mud, worm and deepsea substrate—sink back into the mantle whence they may emerge again. Thus, lava to lava, crust to crust.)

Yet some remnants of life remain in accessible rocks. To the great increase of our knowledge about the planet's living past, rocks preserve a microscopic fraction of earth's organisms for substantial periods. They are transformed in death, of course, and will not bloom or bellow again, but their shapes remain. The rub lies in the not surprising fact that hard material has a significantly better (but still hair-thin) chance of surviving than soft parts. Thus fossil shells abound—relatively speaking—while that which was the flesh of an animal with any kind of skeleton and the entireties of most animals lacking skeletons have vanished. Similarly, young fossils outnumber old ones. Some species, nay entire groups—especially the oldest progenitors—are gone without discoverable traces. Their disappearance (or our failure to find all the relics that survive) create the perceived gaps in the fossil record, which no doubt will always remain fragmentary. But to concentrate on the voids is to ignore the baker's wise doggerel: "As you travel through life, brother, whatever your goal, keep your eye on the donut and not on the hole." Consider the fossils that remain and the reasons they have.

The first requirement for fossilization of an organism is quick, protective burial, usually in soil, gravel, sand, silt or mud. This saves the specimen from being consumed by some passing scavenger; it also protects it from bacterial decomposition and decay through simple oxidation. Once burial has occurred, a number of physical and chemical processes can begin. Most fossils result from one (or more) of the following mechanisms:

An extinct ungulate *Bison crassicornus* in the Ice Age Hall was freeze-dried in Alaska's natural refrigerator. It gave up its moisture by "sublimating" water vapor in the form of ice crystals—a form of natural freeze-drying. The Inca mummy in the Physical Anthropology Hall upstairs survived through the similar mechanics of swift dehydration; relics of extinct human cultures survive as fossils too.

Small organisms, notably insects, may be preserved in amber, the solidifying residue of various plant resins from a source like tree sap. Much of the bug's original substance may decompose, but the carbon of its tissues remains to color the transparent medium and leave a detailed outline. Carbonization also creates many plant fossils, notably of flat leaves pressed between two layers of nascent rock or coal. Volatile substances in the original leaf dissipate, but its carbon particles become concentrated and etch the leaf's outline in the hardening mineral strata.

Most of the dinosaur bones found hereabouts are, essentially, bones. After burial their disintegration slows to a proverbial snail's pace over geologic time. But like wood, porous bone absorbs water, becomes spongy or brittle and very fragile in either case. To preserve finds after excavation, paleontologists handle them with special care, often wrapping them in plaster-soaked cloth and, like first-aiders in the field, splinting them. Back at the Museum, technicians treat them with various hardening agents—chemicals today, beeswax in yore.

Sea snails become fossil shells in several ways. Some are simply buried beyond the reach of bacterial action and oxidation. Others completely decompose, but only after the surrounding mud or silt has hardened around the object to form a cast or impression which remains in the hardening stone. Shells also survive as fossils through "replacement," a refreshingly literal term meaning that silicon dioxide—quartz or natural

Opposite: *Burlesquing the old nursery rhyme, a spider* (top) *and fly* (bottom) *entered a "parlor" of tree sap in the Baltic 38 million years ago. The fluid hardened into amber and preserved carbon ghosts of their remains.*

Overleaf: *This fossil ammonite, an extinct relative of the chambered nautilus (page 351), survived for more than 230 million years because the calcium carbonate of its original shell has been replaced by pyrite, an iron sulfide.*

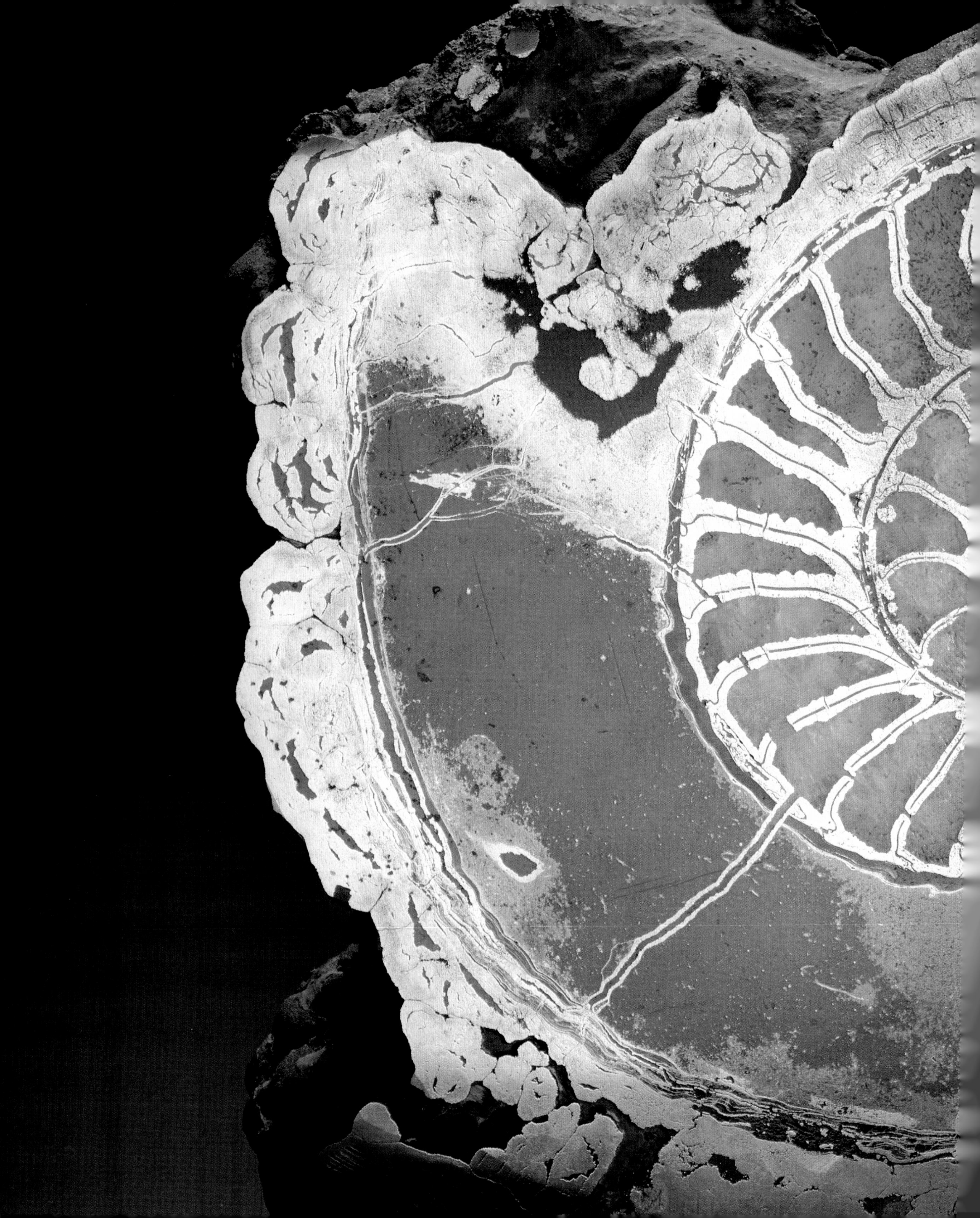

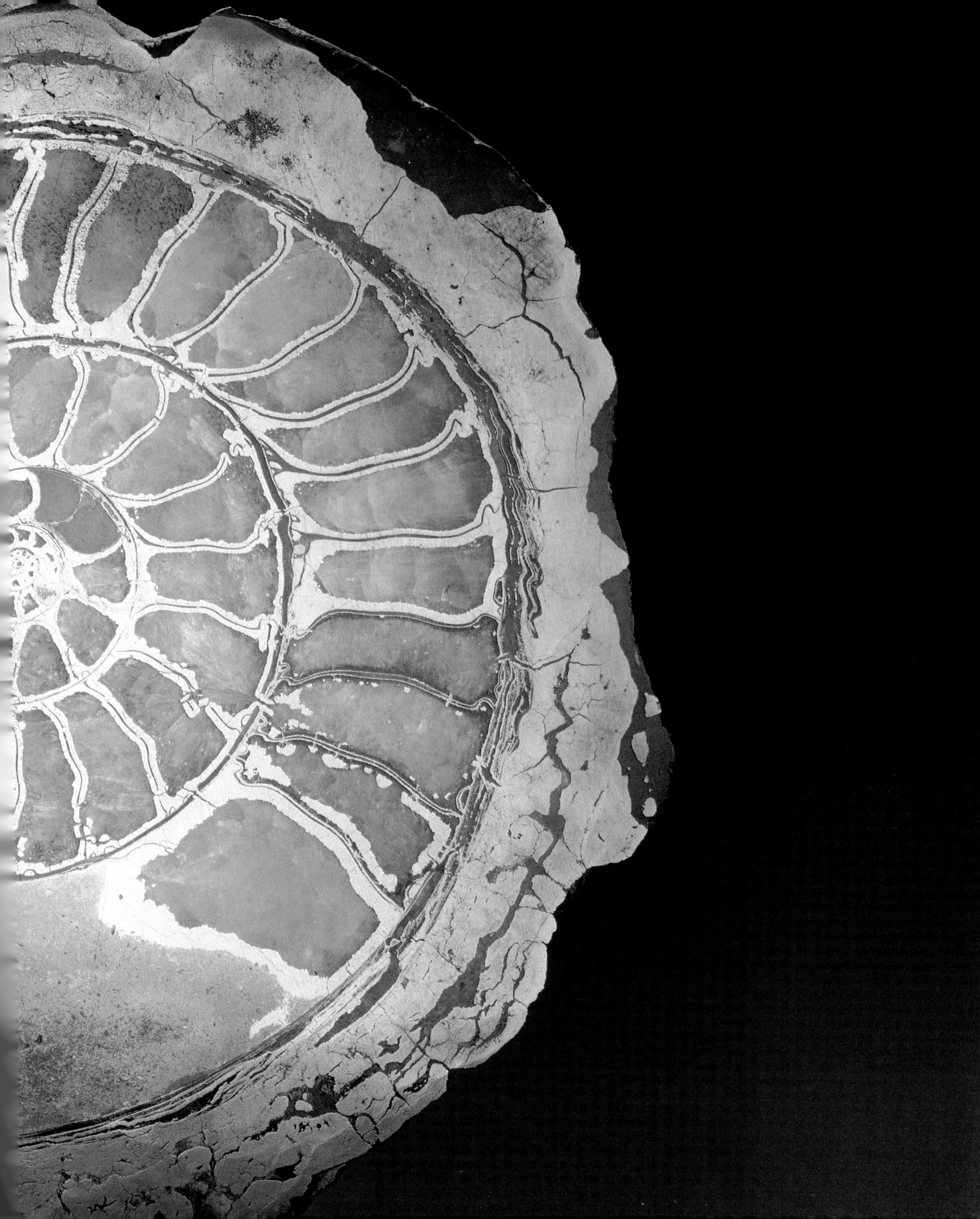

Petrified wood results from "permineralization" or the filling of voids in a tree's tissues with silicas such as agate, jasper and chalcedony.

glass—replaces the animal's material as limestone precipitate around it. The beauty of this process lies in the ease with which the specimen can be chemically extracted from the surrounding material. Placed in a weak acid bath, the calcium carbonate limestone dissolves leaving the insoluble quartz fossil, a replica of the original animal.

Finding that the Glass Mountains of West Texas were made of limestone full of fossils, curators G. Arthur Cooper and Richard H. Grant compiled an unmatched inventory of marine life from the Permian Period, which began nearly 300 million years ago. Their method for recovering wholesale fossil lots seems disarmingly simple. Carving blocks of stone from the mountains like a pair of Yankee farmers mining ice from a winter pond, they shipped them back to the Museum and washed them in acid. A typical 95-pound block produced 6,000 fossils, or nine pounds' worth, and a pound of fine sand. (Processing 40 tons of material this way caused one small problem. The first processing lab was located on an upper floor and one of the tanks sprung a leak. Acid ate its way through the floor before a guard noticed that one of the huge glass light fixtures in an exhibit hall ceiling was full of liquid.) Cooper and Grant's description of a model Permian habitat, published as a massive six-volume monograph, held such significance that the authors won a National Academy of Science gold medal for their work.

Chemical replacement also preserves trees as "petrified wood," while partial replacement or "permineralization" may do the job. Finally the tracks and scat of animals are also fossils in that they constitute remnants of former life, remains that can add information about an animal's behavior just as footprints in sand reveal whether a beachcomber was running straight or browsing. The posture of an animal that drags its tail when walking and raises it to run comes clear in its trail across a muddy riverbank. The disappearance of a track on one side of an ancient lake and reappearance on the other can show that an antique reptile could swim.

So much for the reasons behind fossils. What have they taught us or, to take first things first, what have they failed to show? For one thing, even the oldest fossil found thus far, a clump of bacteria or blue-green algae perhaps some 3.5 billion years old, represents a complex organism. It may

With electric drills like those which dentists use, a team of Museum technicians literally unearths fossils from tons of rock each year. The process requires patience, dexterity and anatomical intuition. It often begins with a block of stone showing only a lump of bone. The trick is to guess which part of which bone of which kind of animal, then judge its position in the rock and proceed with care, because fossilized skeletons are always jumbled. Often—as in the case of this mammal skull—the identity of the animal within the rock becomes apparent only late in the game.

Right: *Some 550 million years ago a shallow-water marine terrace collapsed in a small mudslide, carrying away a sample of Cambrian organisms and burying them beyond the reach of filtered sunlight and oxygenated water. The seven-foot-thick Burgess Shale contained a black-on-black array of rare fossils—some of them unique—with their soft parts remarkably well preserved. Reconstructed in this model, they include: corals, echinoderms, mollusks, worms, arthropods, the earliest known chordate and several enigmas that fit no recognized phyla.*

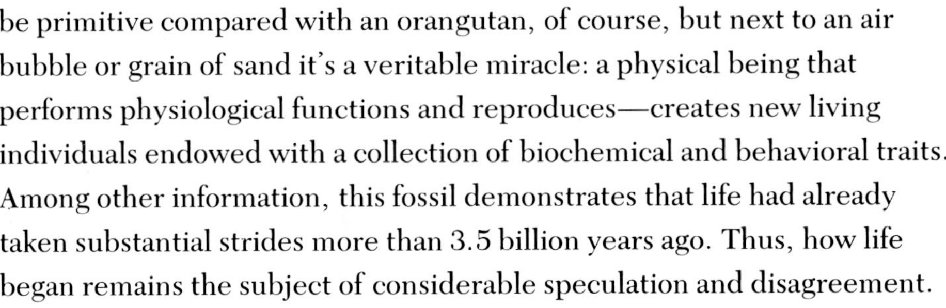

This drill core of a banded 2-billion-year-old Australian rock contains iron oxides—the green, brown and black layers. Dissolved iron, oxidized in the Precambrian seas, settled out as sediment, but whence came the oxygen to combine with the metal? Some paleobiologists believe photosynthetic algae contributed the first oxygen to earth's atmosphere; others believe that solar breakdown of water vapor liberated the gas before life began. New proof of rocky oxidation dates back 3.8 billion years, or 300 million years before the earliest known fossils. This tends to undermine the idea of atmospheric oxygen's organic origin, at least until older organisms are unearthed.

be primitive compared with an orangutan, of course, but next to an air bubble or grain of sand it's a veritable miracle: a physical being that performs physiological functions and reproduces—creates new living individuals endowed with a collection of biochemical and behavioral traits. Among other information, this fossil demonstrates that life had already taken substantial strides more than 3.5 billion years ago. Thus, how life began remains the subject of considerable speculation and disagreement.

For 30 years the most popular hypothesis relied on a spectacular experiment performed at the University of Chicago. A complex, sealed flask was put on the boil. It contained some water to represent the primordial oceans and a mixture of ammonia, methane and hydrogen, the gases most likely to be present in earth's early atmosphere. Above the level of the boiling water a pair of electrodes discharged a spark to simulate lightning. The water boiled, giving off steam in the gassy mix, the electrodes sparked, a condenser cooled the steam to replenish the reservoir whence vapor rose to be energized again in the closed system. This regimen continued for a week and the water turned deep red. It was found to contain an array of amino acids, which are the building blocks of protein chains, which are the components of self-replicating things, i.e., life. Many people leapt delightedly at a seductive conclusion: that the experiment indeed mimicked earth's early atmosphere and that lightning bolts sparked the first organic compounds which collected in oceans or tide pools. Through means that would soon be clarified—or so the optimists hoped—these compounds had taken a quantum organic leap or two and then became living organisms in what Homer had called the "wine dark sea." But problems remained.

As curator Kenneth M. Towe asks: how did the first proto-organisms (let alone the first cells) get along without oxygen? Not that all life consumes it; plenty of anaerobic bacteria still abound. No, the question Towe raised in provocative papers relates to the celebrated ozone layer which screens out the sun's deadly ultraviolet radiation. These rays kill all life as we know it. Thus, mustn't the atmosphere have contained oxygen before life appeared? Yet if one adds any oxygen to the sealed flask in the Chicago apparatus, let alone enough to simulate the reservoir needed for

Dominant during the Cambrian Period, trilobites nearly succumbed twice to other deposit-feeding arthropods during the Ordovician, but rallied through the Silurian until they waned for good during the Devonian. Phacops rana, *found in Ohio, had a mouth like the crustaceans which eclipsed its kind, and a body somewhat like the horseshoe crabs—only distantly related heirs of a common ancestor.*

Opposite: *The longitudinal section of a chambered nautilus shell reveals a unique modern mechanism. The living animal occupies only the outermost chamber, secreting a new wall when it needs more room to grow. One slender organ however, the siphuncle, reaches far back through the tube-like hole in each septum. It controls buoyancy by altering the relative amounts of gas and fluid in the inner chambers. Thus the nautilus rises from benthic depths to atoll lagoons at night, then returns to the abyss with the coming of the light. In earlier times many relatives ranged the seas; now only this one remains in the Pacific.*

an ozone layer, the experiment doesn't work.

Without belaboring all the ins and outs of the various other scenarios, let a particularly intriguing one suffice. Assume that some time after the ammonia-methane-hydrogen stage of atmospheric development, oxygen somehow joined the mix—perhaps through the breakdown of water vapor. Assume further, as many scientists do, that no precursors of life could develop in oxygenated air, though fully formed they could survive once they encountered it. Then introduce them from elsewhere! That hypothesis doesn't rely solely on images of spaceships dispatched by genius cultures manipulating mundane things from some distant planet in another galaxy. Ten years ago meteorite expert Roy Clarke and colleagues at the Geological Survey found formaldehyde in fragments of the Allende meteorite, and amino acids have been reported in a 70-pounder that landed near Murchison, Australia. These clues—or extraterrestrial red herrings—resurrect an old theory of interplanetary micro-organism transport which was soundly refuted since it required the survival of life forms en route through space. Now, instead of talking about spores or viruses, scientists hypothesize the meteoritic arrival of "a precursor for carbohydrates." A century ago Darwin had opined: "It is mere rubbish thinking at present of the origin of life; one might as well think of the origin of matter." Since his day, science has gained the theoretical understanding to contemplate even the latter, though no easy answers are yet forthcoming. Like earthly habitats, times change.

Cardiospermum coloradensis, *found in the Green River Shale, lived 50 million years ago and survives in warm New World climates as a "balloon vine."*

Opposite: *Once spore-bearing plants colonized the land, they developed increasingly adroit ways to claim dominion—or lost out. One early group carried a single spore on the end of its stem. Not enough, it perished. Another group, the trimerophytes, produced spore sacs on the ends of complex side branches, continued to grow and loosed many more potential young to the wind. This lycophyte, a species of* Baragwanathia, *took the middle road 380 million years ago and developed spore sacs where its spiraling leaves joined the stem. Though bested by the trimerophytes—which became ferns—and all the seed-bearing plants which in turn became every living tree, grass and flower, the lycophytes still survive as club mosses and quillworts.*

If the question of vital origins remains unanswered at the Museum (and everywhere else), much has been learned about what happened once life got started here more than 3.5 billion years ago. Frustratingly, most of the information illuminates only the last 570 million years or so. A large gap lies between the forming of the first known fossil and the time when fossilization became common. Three billion years of biological history passed leaving only occasional traces. It is as if an amnesiac, aged 44, sees his own picture in an out-of-town paper along with the news item that his original persona has just been declared legally dead. While he still can't remember much about what happened before he disappeared seven years earlier, he knows he's very much alive. By the same token, fully 84 percent of the organic past remains mysteriously obscure to us today.

By the start of the Cambrian Period (which lasted from 570 to 500 million years ago), the seas had become alive. In all probability every phylum of marine organism known today had gained a metaphorical foothold, including all the nominal worms. Oxygen-dependent animals appeared in such numbers that the photosynthetic revolution was all over but the shouting. Trilobites (now extinct) and brachiopods (scarce "lamp shells" today) dominated the sea floors. Primitive sponges, snails, clams, crabs and cephalopods (now squid and chambered nautilus) were diversifying. The first chordates were on their way to becoming primitive vertebrates in the form of jawless fishes. During the Ordovician Period (500 to 435 million years ago) the fish gained jaws, fins and varied forms. Starfish, echinoderms, corals and crinoids diversified along with mosslike bryzoans, the first "amphibious" plants that could colonize the barren land where nothing lived.

That invasion, launched onto the beachheads of riverbanks and tidal flats, became successful during the Silurian Period (435 to 395 million years ago). For plants to survive out of the supportive, wet and permeating sea, they had to develop thicker cell walls for support and moisture conservation. They needed to invent vascular systems to deliver nutrients, specialized root parts and pores for transpiration. Invertebrate animals quickly followed the pioneering plants ashore, small scorpions and their heirless nine-foot-long eurypterid relatives. They could make this explora-

tion with relative ease because they were fully ambulatory under water and equipped with shells that could support their weight on land. Further, their book gills allowed them to breathe air so long as these structures remained moist.

During the Devonian Period ("the age of fishes," from 395 to 345 million years ago) other arthropods—their ancestral forms now lost—adapted to terrestrial life, and became plant-eating insects in the process. The relatively empty environment did not remain predator-free for very long because by mid-Period lobe-finned fishes were waddling across the mudflats and giving rise to the first insect-eating amphibians. These animals remained dependent on bodies of water during two stages in their lives at least. They deposited their membranous eggs in water and the larvae, born in water, developed there until their metamorphosis into adults.

The major botanical groups, which lacked both true seeds and flowers, also depended on wet habitats for reproduction. As ferns do today, each increasingly large seedless plant had two forms: as a sporophyte it released spores which became leafless gametophytes. These small, intergenerational plants then produced sperm and egg cells which could combine to become sporophytes; but the sperm still required water in order to swim about on their inseminating missions. This chancy business worked well enough in wet climates but left considerable room for less climatically dependent reproduction and greater metabolic economy. Its shortcomings aside, five major plant groups employed the spore mechanism. Two of them would die out completely; two others led to the club mosses (represented by about 150 species today) and the scouring rushes or horsetails (about 60 species now). The most successful group, the trimerophytes, developed the most highly ramified branches and most numerous spore-producing organs. In time they would give rise to the ferns, represented by 300 species now, and all the 350,000 species of seed-bearing plants that cover the earth. Those prolific seedless plants also discovered the advantages of height during the Devonian. The tree that raised its crown above its neighbors won the prize of more sunlight for photosynthesis; the race went to the thickest-crowned and the tallest.

Independence from external water for reproduction came during the

Opposite: *Because of their small size and fragility, relatively few insects turn up as fossils. Fortunately the Green River Formation preserved upwards of 300 species including the cricket* (top) *and robber fly* (middle)*. Barely an inch long and very rare, this fossil scorpion* (bottom) *was one of the first invertebrates to venture onto land. It came ashore in what is now New York State 405 million years ago.*

Opposite, top: *Another Green River denizen,* Mioplosus labracoides, *lived 50 million years ago during the early Eocene in what became Wyoming. Its closest modern relatives are yellow perch.*

Opposite, bottom: *The oldest fossil material found thus far—the reddish mound in the center of this specimen—comes from western Australia. Sliced vertically, its layers reveal stromatolites, a sedimentary structure influenced by algae. Algae evidently grew in still water, survived paper-thin burial, then grew in a new layer which in turn was thinly buried again and the process repeated. The organisms themselves have long since decomposed and been replaced by silica minerals which preserve signs of their activity 3.5 billion years ago.*

Carboniferous Period (345 to 280 million years ago). Descending from amphibians, reptiles achieved independence with the hard-shelled egg which contained a wet microhabitat for the developing embryo. Among plants, the primitive gymnosperms—conifers, cycads and ginkgoes—produced naked seeds almost as their ancestors had produced spores, only without the special and distinct intermediate generation. Male pollen grains released to the wind landed on cones and migrated down a moist tube to the ovary where fertilization took place. The new proto-organism then formed a sizeable protective structure and was released to the wind. This process increased the reproductive success of these plants which began to eclipse the less efficient seedless plants.

The Permian Period (280 to 225 million years ago), most notable for its negative impact, ended the Paleozoic ("old life") Era. For reasons that elude explanation, fully half the orders of marine invertebrates became extinct, leaving myriad niches empty—but not for long. The Mesozoic ("middle life") Era began with the Triassic Period (225 to 195 million years ago). Cycads and conifers dominated the landscape while beneath their canopies the reptiles diversified from small forms rather like today's lizards. The first dinosaurs arose from the "stem reptiles" (cotylosaurs), so called because they produced the reptiles that now survive. This group would also produce birds. About the same time cotylosaurs gained prominence, the mammal-like reptiles (synapsids) appeared, and remained judiciously inconspicuous. For the dinosaurs, bigger meant better as the Mesozoic Era continued. They grew to legendary sizes, a varied group whose members filled every niche that could be muscled big enough. They evidently ate everything from marsh grass to each other. Meanwhile, groups of small shrewish mammals scurried about underfoot, too small for their cousins' attention, invaded the trees and watched for their next adaptive opportunity.

The Cretaceous Period witnessed a couple of major trends. Flowering plants (angiosperms) developed and started becoming fabulously successful thanks to two mechanisms. One, the seed capsule was protected within the mature plant, where fertilization was instigated by animals, notably insects, instead of the uncertain wind. These ubiquitous pollinators, in

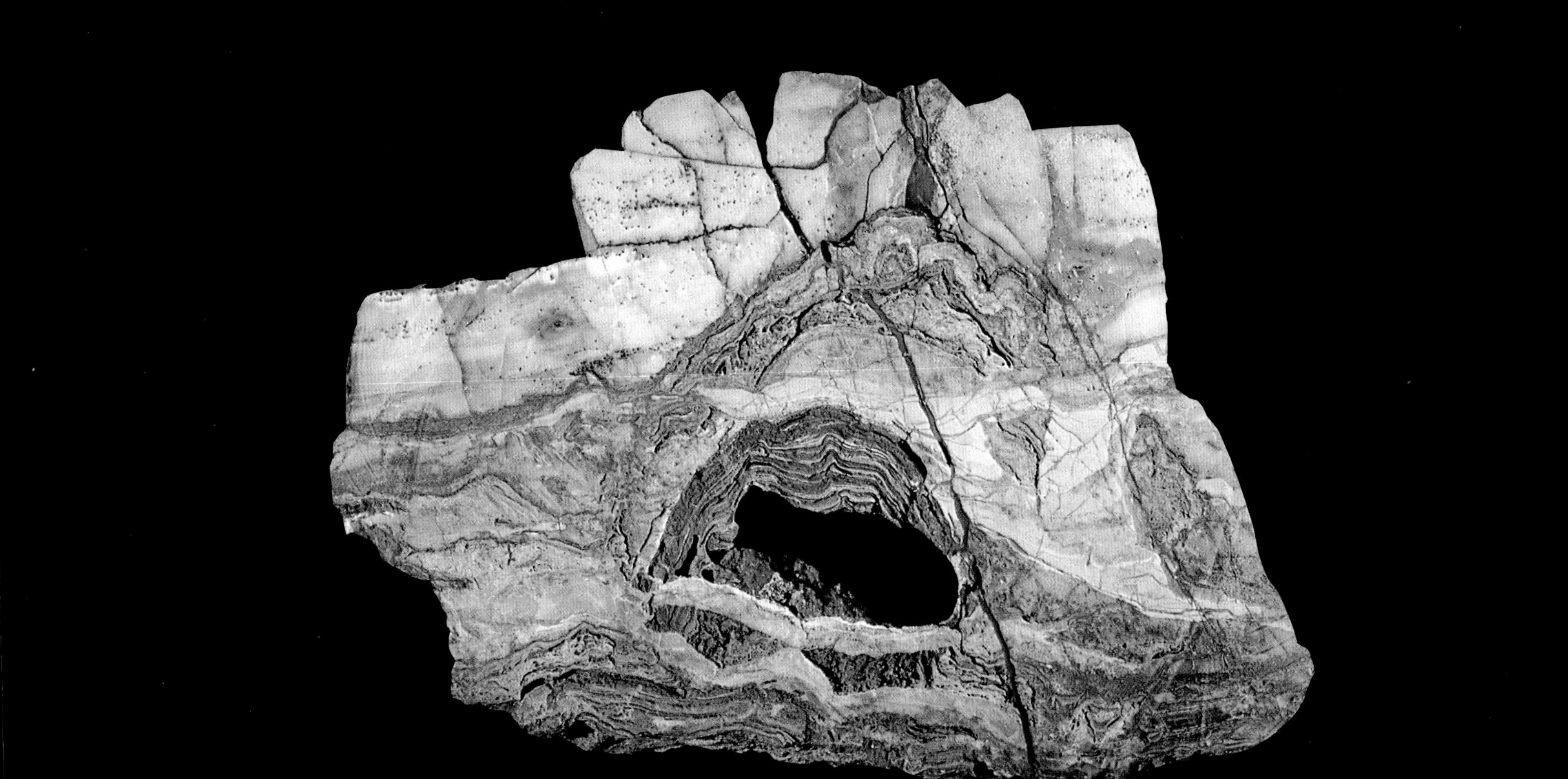

Back to
the Water
The
Tree

selfish pursuit of food, went from flower to flower, happenstancially inseminating as they went. (In time many plant and insect species would adapt specifically to suit each other in absolutely interdependent relationships. Without the plants, the insect had no food source at some stage in its life; without the insect to transport the pollen, the plant could not generate heirs.) Two, the angiosperms had found a way to avoid, in essence, the gymnosperms' wasteful business of providing a food source and protected capsule for each embryonic seed regardless of its likely future. The newly evolved pollen grain contained two nuclei: one to fertilize the egg, the second to trigger the growth of embryonic food. If a flower were not pollinated, the parental plant saved the energy of developing it further and only expended valuable physiological resources on fruiting elements with proven potential. This saved enormous amounts of caloric energy which the conifers and cycads spent on every seed and its oily or waxy capsule. (In unpardonably anthropomorphic terms, this compares with the relative economy of two mothers with teenage daughters at a summer day camp. One mother runs up large grocer bills and packs a daily lunch for her dieting child—though it's usually wasted. The other mother opens a charge account at the local snack bar and only pays for food that's actually consumed.) The angiosperms, performing less metabolic work prior to reproduction, produced new generations with botanically unprecedented speed on land—in weeks instead of at least a year for each crop of potential offspring. The competitive advantage over older plant groups soon became abundantly clear, as flowering plants covered the earth.

The Cretaceous Period also saw the widest diversification of the dinosaurs who ruled the terrestrial roost for 70 million years. Reptiles which held their bodies directly above their legs, many dinosaurs became functionally bipedal, carrying themselves across the ground like running birds. (The others walked on four laterally straight legs—not splayed ones like lizards, crocodiles, turtles and the like. By definition, true dinosaurs also excluded such nonterrestrial relatives as the fabulously serpentine marine reptiles and the flying pterosaurs.)

Some paleozoologists attribute the cool-blooded dinosaurs' increasing

Opposite: *The first genera of trees included* Callixylon, *which grew 100 feet tall in the late Devonian Period 350 million years ago, and gave rise to gymnosperms—the conifers and such. This specimen from Kentucky fell, rotted partly and was nearly flattened before becoming fossilized to be preserved, found and raised again. In life its relative height was an asset. Taller trees received more sunlight than their neighbors, more energy for both growth and reproduction. Plants with broader leafy crowns—more area for photosynthesis—did the same, so true trees came to eclipse the ferns by literally rising above them.*

One face of the Time Column sketches the evolution of life on earth over 700 million years in a single monolithic illustration. Starting at the bottom left: live one-celled orbs begin to congregate into colonies and complex organisms. Jellyfish appear, then other invertebrates such as corals and trilobites, then squid-like mollusks and huge arthropods in the clawed sea scorpion. While plants colonize the land, fishes flourish in the sea, the first vertebrates to dominate a habitat. As insects, the arthro-

size to a thermal strategy like the one that still works for warm-blooded whales in frigid oceans. These dominant reptiles may never have gone to the caloric expense of becoming "endothermic," Dr. Hotton believes, and thus kept their nutritional requirements substantially below those of birds and mammals. Living seasonally in regions where the sun warmed them, they also generated some heat through simple exercise as they constantly lumbered about. (Even the carnivores needn't have been particularly agile or swift in absolute terms—only a little faster than their prey.) They grew increasingly large because greater size means less heat loss as the body's surface-to-volume ratio declines. From one perspective, their phenomenal growth compensated for the lack of warmth produced by the mammals' specially evolving internal organs. Always on the move, dinosaurs lumbered as far north as arctic Alaska during the long days of summer and back to warmer regions for the winter, in annual treks of many hundred miles. Discovery of their remains in the higher latitudes led other paleontologists to hypothesize that dinosaurs were warm-blooded and behaved like frantic mammals madly pursuing large numbers of prey to maintain a high-calorie metabolism. That portrait has not been widely accepted after its splashy debut. Just because some individuals died on the Bering Sea coast doesn't mean they lived there year round; people die in Atlantic City's casinos who haven't spent their lives there. To test the warm-blooded hypothesis, other paleozoologists have tried to mount a few specimens in the mammal-like postures necessary for the hyperactive life-style depicted in some speculative paintings. They found the bones wouldn't fit in ways that made orthopedic or anatomical sense.

The Mesozoic ended like the previous era, in what amounted to disaster for dominant animal groups. In a relatively brief period—no longer than 10 million years according to one estimate—the dinosaurs became extinct along with many other reptiles. Theoretical explanations suggest a wide range of possible reasons. The most dramatic scenario, and in many ways the most captivating one, features a huge meteorite which struck earth and enveloped it in a cloud of debris that blocked out the sun for a few years. That would have meant the death of green plants, then the death of the great herbivores and the passing of their predators. But major

pods move ashore. Amphibians follow as the carboniferous forests rise (top left and bottom right), then produce reptiles which radiate as dinosaurs, birds, rodent-like mammals. When the giant reptiles die out, mammals diversify into a company that includes primates, then early man dwarfed by tusked and toothed members of the family. Possessing a single biological asset, their expanded brain, the distinguished races of modern man survey the entire orb of earth.

groups of smaller, warm-blooded animals were not affected, suggesting that the dinosaurs may simply have pushed the physical and physiological limits, or perished when the climate cooled slightly. The meteorite theory has been contested by paleobotanists on the basis of vegetative evidence. They say the same flora grew on either side of the Mesozoic/Cenozoic boundary when the "new" fauna begin to appear. As so often happens, the time zones established by paleozoologists and geologists have little meaning for their botanical colleagues, a group of specialists only recently grown large enough to acquire a real disciplinary identity. Paleobotany curator Leo J. Hickey says that if his colleagues had their way the ancient calendar would feature a "Paleophytic Era," ending after the first third of the Devonian Period when the spore-bearing plants waned, a "Metaphytic Era" marking the age of the gymnosperms into the Cretaceous, and so on. This proposal, not altogether capricious, points out an important phenomenon.

Throughout biological history, plants, or plantlike organisms, have been the pioneers and animals the opportunistic camp followers. When the land was utterly barren it took an autotrophe—an organism that manufactures its own food from inorganic elements, a plant by one definition—to make the first move. Once a self-starting food source was there, the heterotrophes—organisms requiring ready-made organic food whether meat or vegetable—could follow. Once herbiferous insects occupied the new botanical habitat, they then provided the nutritional wherewithal for other insects and higher animals which consumed them and then learned to consume each other. The story throughout is one of exuberant energy, profligate activity and wild opportunism.

These universal traits of living things are part and parcel of evolution, Charles Darwin's momentous discovery. Through it he made a contribution of inestimable value to our understanding of this living world. But he underestimated the extraordinary and irresistible dynamics of all biological creativity. For one thing, his tautological euphemism "survival of the fittest" glossed over the constant of a 100 percent death rate in the course of time. For another, he emphasized the apparently simple matter of "descent," implying a fairly simple, linear process. Successive generations

of scientists have been fleshing out the picture he sketched.

Consider an essential meaning of the well-chosen word "radiation" which is often overlooked, unseen for the wood of the "family tree" metaphor. Any multigenerational arrangement, from Princess Diana's pedigree to the diversification of the synapsids, conventionally appears in two dimensions on a page. In actuality every "family tree," whether a Windsor's or a wombat's, grows in a manner better represented by a circle. The reason: descendents do radiate in a literal sense and a figurative one. In human terms, Mom and Pop raised a family in downtown St. Louis; their kids moved to the new suburbs ringing the city; their children in turn scattered to New York, Chicago, Houston and California, save for a couple who returned downtown. So be it with biological species. Geographically they tend to expand any original range as far as they can. Most important, they also randomly alter their traits in an all-encompassing manner through genetic mutation. One isolated group may appear to grow in only one metaphoric "direction," as the dinosaurs grew bigger. But in a more comprehensive context, dinosaurs belonged to a tribe that had already grown smaller in the proto-mammals, higher in the flying reptiles, deeper in the marine pioneers.

For illustration's sake, consider a model viable species to be represented in a single, perfectly shaped apple tree perched on the top of a conical hill. With the end of summer the tree drops its evenly distributed fruit. Some apples get caught in the grass beneath the perfectly circular canopy; some roll down the hill in every direction of the compass. Some are eaten by browsing deer, some by birds and worms. (Permit me the fiction that no seeds survive digestion.) Some others remain intact near the parent tree and their seeds turn to seedlings.

By definition most of the apples are "normal." They neither display nor possess any new trait that their offspring might inherit. They contain no spontaneous genetic changes—mutations determined by typographical errors in the genetic code, whether insertions, transpositions or omissions. Some of the apples, however, inevitably do possess aberrant mutations. Such is life. A few lack sugar and taste terrible, even according to a crow's dubious culinary standards. A few never turn red but remain green. A

Opposite: Thrinaxodon liorhinus, *an evolutionary link familiarly known as "Baby Doll," lived in South Africa during the early Triassic Period about 225 million years ago. This most mammalian of the synapsids or "mammal-like reptiles" gave rise to modern reptiles while its reptilian heirs all died out.* Thrinaxodon *has also been found in Antarctica to support the theory that the polar continent abutted Africa during the Mesozoic Era. Still other fossils reveal that adults of the species protected their young, a trait shared by most mammals but by few reptiles today.*

Opposite: *A single ginkgo tree may have produced these variously shaped leaves, preserved for 75 million years in the Fox Hills sandstone formation of the northern Rockies. Related to conifers and once nearly ubiquitous, ginkgoes became extinct in North America during the Miocene epoch when mountain building occurred in earnest 20 million years ago. Some species survived in the Orient to become temple trees treasured for the edible seeds within odoriferous fruits. Imported as an ornamental to the United States from China, the modern* Ginkgo biloba *will often grow shoots bearing diversely shaped leaves like these fossils rather than the mature branches' two-lobed leaves. Each form may mimic some ancient ancestor as genetic memory expresses itself in morphological capitulation.*

few have abnormally waxy skins. These mutants are also evenly distributed around the circular canopy atop the perfect hill. A passing deer bites a sour apple and spits it out in favor of the normally sweet ones which abound. Green ones, camouflaged in the grass, escape the deer's notice. Waxy ones don't rot fast; one rolls down the hill to reach a stream and fetches up on the bank of a pond some miles away. All of these survivors pass on their new traits to their offspring—to the latter's benefit or detriment, depending on circumstances.

Meanwhile the waxy apples that didn't roll downhill get eaten by the deer. Sour ones and green ones that roll down the hill rot in the stream and feed crayfish in the pond without taking root anywhere. Thus many of the apples with new traits don't survive to endow a new strain of trees. Meanwhile, a higher number of normal apples also take root. Any one of them has slim enough chances, but because there are so many more "normal" individuals carrying well-transmitted codes, more of these survive through the normal course of events. Thus the next generation of apple trees on the hill is dominated by ordinary heirs. A few with mutational advantages colonize new environments such as the distant lakeshore.

However, when some major change occurs in the region, stress occurs. The normal trees may suffer most. An abnormally cold winter kills some trees; a string of frigid winters kills all the normal ones, leaving only trees with the mutation—by definition a genetically determined abnormality—to survive cold.

In species like horseshoe crabs and ginkgo trees—so-called "living fossils," survivors of very ancient times at any rate—all offspring with aberrant traits died out over the millennia. Change can be fatal. Only those crabs and ginkgoes with the "normal" traits lasted through the many environmental upheavals of earth's history. Among advanced dinosaurs, all died out leaving no heirs. Among the mammal-like reptiles, on the other hand, diversifying offspring survived while the less specialized originals lost out in the biological competition. As idiomatic Sanskrit has become a dead language, the early reptiles constitute a dead group. But they have living survivors—markedly changed—as Sanskrit survives in Hindi, Urdu, Romany and Punjabi.

Preceding pages: *Extreme anatomical eccentricity can have a high price. Over the course of 1 million years,* Megaloceros giganteus *developed antlers that spread 12 feet and weighed 100 pounds, the largest racks ever seen. The animal ranged across Eurasia from China to pre-insular Britain and even reached North America. Well adapted to grassy steppes and arid tundra, the Irish Elk did not function well in forests, which spread northward after glacial retreat. In addition, predatory man probably hastened its extinction 11,000 years ago. Frequently preserved in ancient bogs, the skulls are still found in noble British houses mounted like paleolithic hunting trophies.*

Opposite: Smilodon californicus *frequently appears in the La Brea tar pits of Los Angeles. Its generic name does not punnily refer to a predatory grin but to the root words for knife or chisel and the sabertooth's habit of stabbing with its fangs. Several independent lines of mammalian cats and their marsupial counterparts developed the same winning combination of huge upper canines and a lower jaw that opened wide enough to accommodate them in action. Each group depended on brute strength and heavy weaponry to kill ungainly herbivores during the Pleistocene. But having traded away speed and agility in the evolutionary process, they could not survive the passing of big, slow prey species.*

In the Las Vegas idiom, it's a crapshoot. Darwin was right in one monumental respect: evolution occurs. By that I mean that species change over time. He was wrong in one major particular: that "natural selection" provides the prime mechanism for change. It presupposes only a kind of lucky and passive survival apparatus, not the dramatic, intrinsically creative appearance of new traits that might or might not suit new conditions and habitats. When a new trait proved pragmatically useful and its few accidental possessors survived to mate, it could supersede older traits in the offspring.

Because mutation can occur anywhere in the genes, it becomes manifest on a random basis that might be represented by a full circle of 360 degrees, not a branching line. In a viable species, any departure from the norm could be immediately detrimental for the individual; only new color variations occurring within 5 degrees of an artist's color wheel, for example, might be advantageous. But when conditions change and a species undergoes stress, a new color pattern can make the difference between survival and extinction. The classic example here (and in the Evolution Hall) involves Britain's peppered moth, which has mottled wings. For millennia these moths spent daylight hours sitting on grayish lichens with their wings extended. Because the lichens were light-colored, the light-colored moths thrived, though for as long as anyone noticed about 1 percent of each generation had darkly mottled wings. Few of these survived because birds saw and ate them. But with the coming of the Industrial Revolution, smoke from Dickensian factories killed the delicate lichens and blackened tree trunks and rocks. More and more of the dark-colored minority of peppered moths survived to breed while the light majority became newly vulnerable and turned into birdfeed. In the course of a century, the dark-colored peppered moth became typical. But the genetic code still dictated variations and produced some moths with light-colored wings. When Britain began clearing the air a couple of decades ago through environmental legislation, lichens returned and more of the new minority began to survive.

Darwin may also have erred in believing that evolution only happened very gradually. Neo-Darwinists, called "gradualists" in the trade,

Preceding pages: *Vertebrates have been around less than one-tenth of earthly time and birds (for example) about one-twenty-fifth, but evolution has produced an incalculable variety of living things through the mechanisms of variation and mutation. Whether the latest flu virus that wreaks pandemic havoc among mankind, or this Indonesian argus pheasant which avoids men, every organism has a primeval inheritance and an unpredictable posterity.*

have tried to fit the master's theory to new data as slowly revealed in the expanding fossil record. But gradualism doesn't seem to work in the opinion of a growing new school which champions Punctuated Equilibrium. In a nutshell the "punctualists" read the fossil record this way: an ecosystem settles into a state of equilibrium; the fossil inventory in a given place often shows the same relative numbers of different species over long periods of time; the same plants, herbivores, predators, etc., appear in several rock strata. Then quite abruptly a new stratum, representing a new time, displays a different array of plant and animal fossils, radically changed descendents of the previous groups but often with perplexingly few intermediate forms. Herein lies the "punctuation" or a great leap forward if you will.

A single example of this would hardly amend Darwinian law, since it might represent a local climatic anomaly or a simple break in the record. But spurts of evolutionary change seem more common than exceptional throughout biological history. This leads the "punctualists" to suggest the new ingredient in the recipe of creative evolution: stress. Given the long-term maintenance of a set of environmental and ecological conditions, all a habitat's plants and animals settle into a modus vivendi. So many insects eat so many leaves; so many trees die but so many remain to bear fruit; so many birds live off the bugs and enrich the ground with their droppings that fertilize the trees; and so forth. But when conditions change—for climatic, tectonic or inexplicable reasons—the modus vivendi collapses as quickly as the big-car market when Arabs raised oil prices. Some of the species perfectly adapted to previous conditions are no longer best suited to the new conditions; but if some of their mutant offspring are well suited, then the line continues. Assuming these new conditions last a few million years, these offspring multiply, occupy environmental niches vacated by some of the losers and become thriving species—new species.

That's what happened when the dinosaurs died out, whether they were exterminated by a meteorite's dust cloud, by aberrant radiation or any other cause such as simply growing too large. Small mammals, waiting in the wings, swiftly adapted to play new roles—often ones of stellar ferocity previously played by the dinosaurs. Why? Because the random mutations of each generation allowed a few individuals to take advantage of newly available opportunities to act predatory, for example, and openly chase others across the land. The emerging mammals did so well in the early Cenozoic Era that more and more of them possessed predatory traits. As they became better fed and more robust, they found other mates who shared the new traits and passed them on. The first adventurers happened to appear more "fit" than their normal, timid, inconspicuous cousins. But they were made that way by an accidental change in their genetic code—a code that dictates a few changes of every conceivable sort in every generation.

It deserves special emphasis: a comprehensive collection of mutations can probably be found in every generation of every species. Throughout its existence each species stands ready to seize the changing day.

A MUSEUM OF TAXONOMY

The Naming of Cats is a difficult matter,
    It isn't just one of your holiday games;
You may think at first I'm as mad as a hatter
When I tell you, a cat must have THREE DIFFERENT NAMES.
—T.S. Eliot

A library of rocks and fossils, the Museum is also a library of living species in the form of their carefully preserved remains: bones in boxes, bivalves in drawers, plants on paper leaves, needlefishes in alcohol, frogs in formaldehyde, beetles on pins. People here spend much of their energies doing what librarians do—cataloguing—but with an added element that eclipses the rest. The work involves more than maintaining card files on 60 million items, assigning them slots on shelves and keeping track of their whereabouts among borrowers. It centers on original work: the identification, naming and classification of biological species represented by the collections. These functions, collectively called taxonomy or systematics, underlie most of the Museum's research in several biological sciences, even touching on the science of man. Taxonomy is one of the Museum's longest suits, oldest preoccupations and most intrinsically taxing disciplines.

It is also one of the most widely misunderstood and least appreciated by outsiders (whether other scientists or laymen), a subject that causes tediously repetitive intramural debates varied by sly rhetorical tangents and bits of frank malice. One geologist here, for instance, believes the entire taxonomic club isn't worth the few rocks it would take to stone its members to death. What galls him most: their devotion to "higher systematics," the grouping of organisms into categories based on perceived relationships when both the perceptions and the criteria are fluid. "It's all subjective," the geologist sneers. "They write everything up, then tear it all apart and write it up differently." Accustomed to slurs, these specially disciplined and motivated scientists know that biology couldn't get along without them, nor without this work which the Smithsonian has performed for more than a century.

This "snout beetle" could be a candidate for introduction to North America, where more than a dozen other exotic weevils already help control weeds. Or, if accidentally introduced here, it could become a pest. Taxonomists have named some 50,000 species—most of them innocuous—in this group of vegetarians. Perhaps five or ten times that number remain to be identified as taxonomy pursues its goals of identifying and classifying the world's organisms.

Overleaf: *These insects have two things in common. All beetles, each possesses a pair of protective wings to cover the functional pair when not flying. Secondly the hard, horny protective wings often display extraordinary beauty and these specimens were selected from the collections simply because they look spectacular. Chosen for the "Splendors of Nature" exhibit, they come variously from Java, Panama, India, Thailand, East Africa, Texas, Liberia, Argentina, Arizona, Mexico, Zaire, Borneo, China, Cameroon, South Africa and the Philippines.*

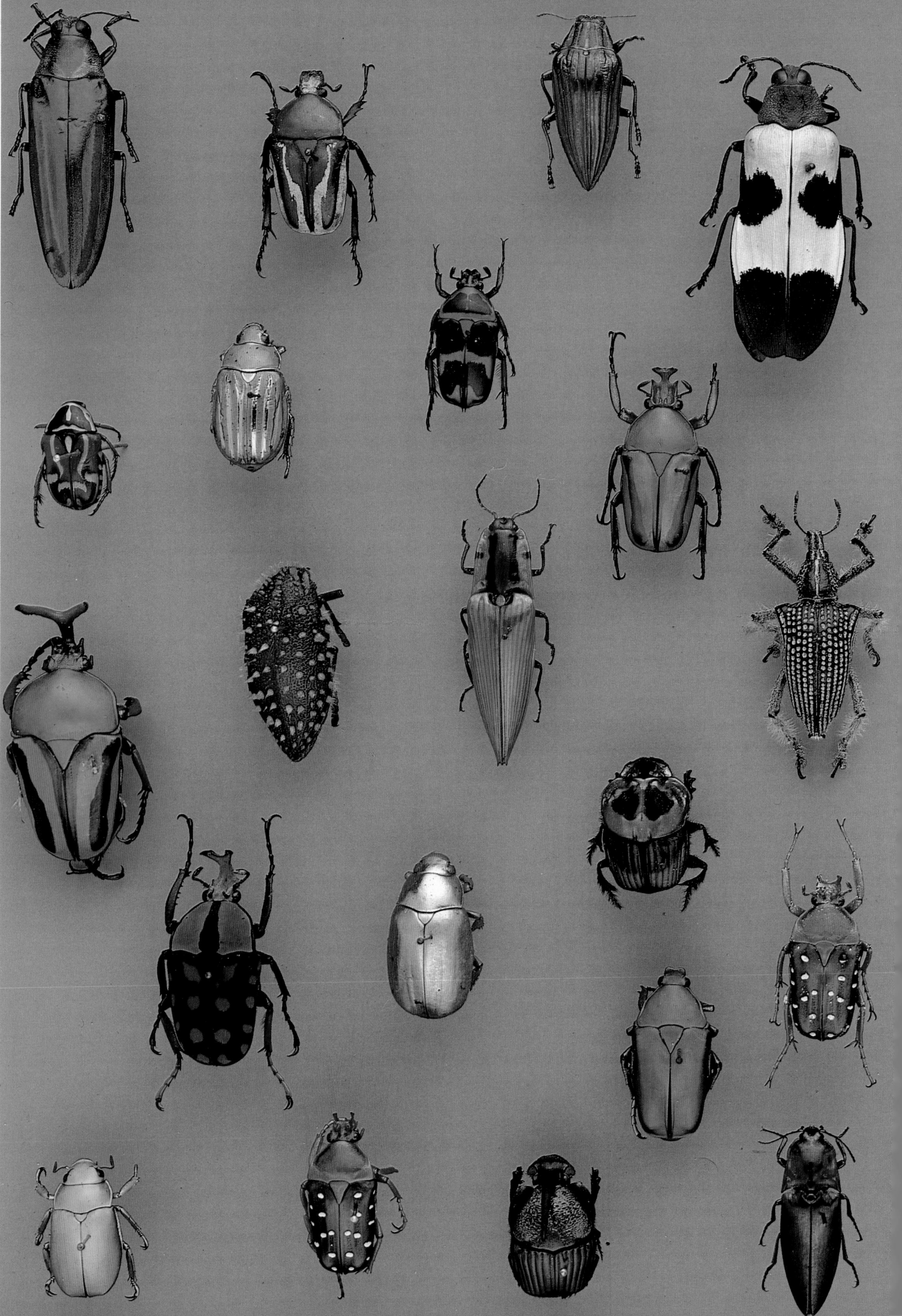

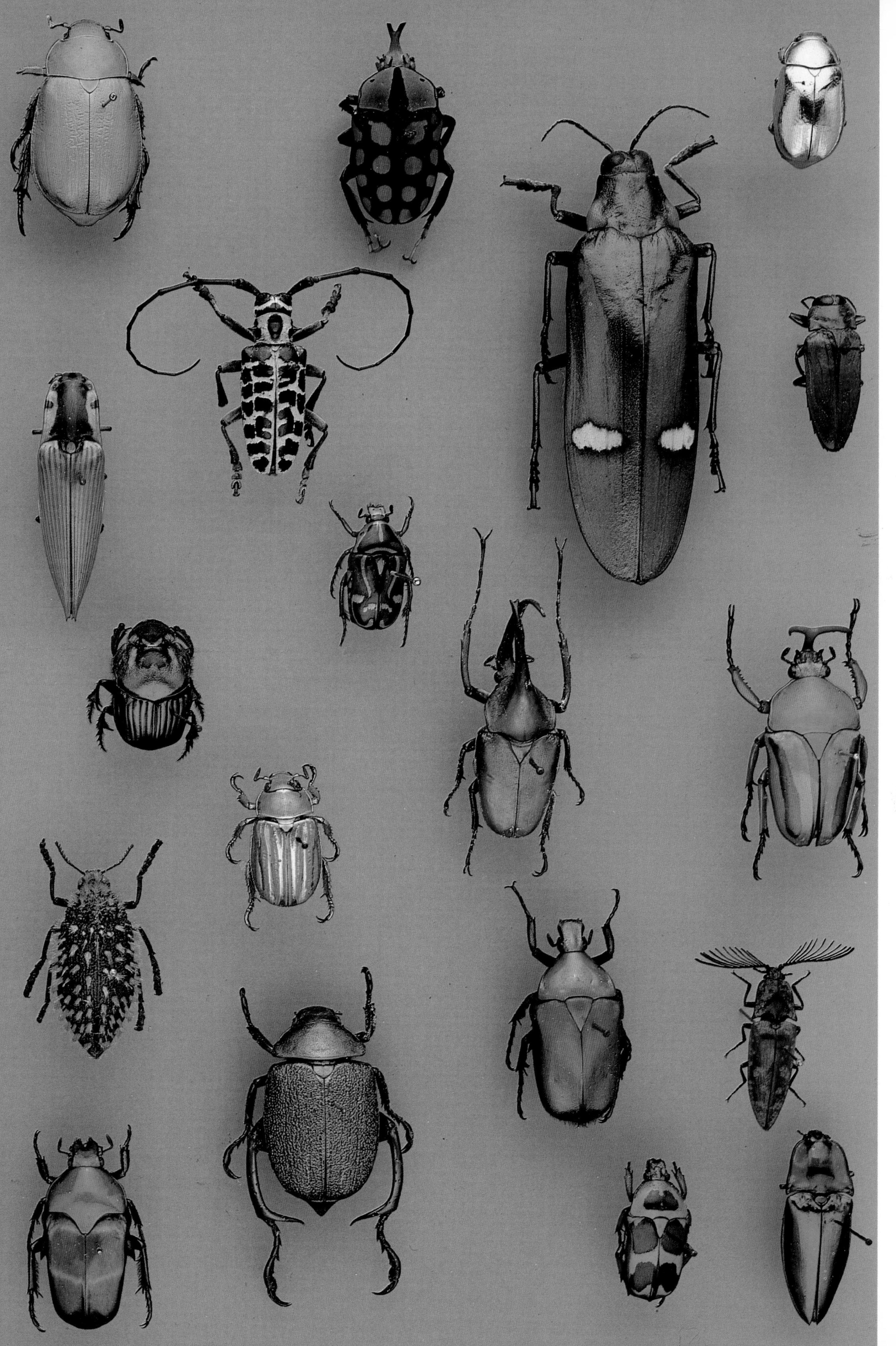

Opposite: *One of the "longhorn beetles," so called for their exceptionally long antennae, the harlequin beetle inhabits tropical forests. Using its 7-inch-long forelegs for climbing, this male of* Acrocinus longimanus *lives on sap of several kinds of trees but especially that of the fig trees. The larvae of these beetles live as borers beneath the bark and in the heartwood of injured or dying trees. The mature larvae, which reach a length of 5½ inches, are sometimes roasted and eaten by jungle-dwelling natives.*

As the Institution grew and its bureaus multiplied, taxonomic pursuits coalesced in the Museum, which became—and remains according to informed, independent opinion—one of the very few "most important centers of systematic biology in the world." That makes this building undramatically and crucially the Springfield, Massachusetts, and Akron, Ohio, of the life sciences. Springfield? The home of G. & C. Merriam, of course, publishers of Webster's dictionaries. Because clearly defined and understood species are to biology what words are to literature, namely the basic elements of the larger discipline. Akron, tire capital of the world? Yes, because without its stock-in-trade, crucial systems we all depend on would collapse; medicine might take several steps backwards, agriculture would become the playground of uncontrollable pests, ecology a guessing game and research biology for any purpose a charade.

"Taxonomy is at the same time the most elementary and the most inclusive part of zoology," wrote the distinguished mammalogist George Gaylord Simpson a generation ago. It is "most elementary because animals cannot be discussed or treated in a scientific way until some taxonomy has been achieved." It is "most inclusive because taxonomy in its various guises and branches eventually gathers together, utilizes, summarizes and implements everything that is known about animals, whether morphological, physiological, or ecological."

It begins with the delicate business of individual species identification, "alpha taxonomy" in the trade. It then proceeds through "beta" and "gamma taxonomy," most often called "systematics," the discovery and state-of-the-art definition of kinship connections between groups of organisms. The eventual goal is to reveal how all organisms are related to each other, an achievement we might expect the first Thursday after the Second Coming. That goal implies the illumination of each step in biological history and the charting of a comprehensive family tree detailing all the world's animals, plants and protists (most elementary unicellular organisms), living, dead or in undiscovered states of suspended vitality. Before such a grand design can be drawn, however, an incalculable number of unknown details must be defined, then refined.

For now a perfect, comprehensive and detailed genealogy of life on

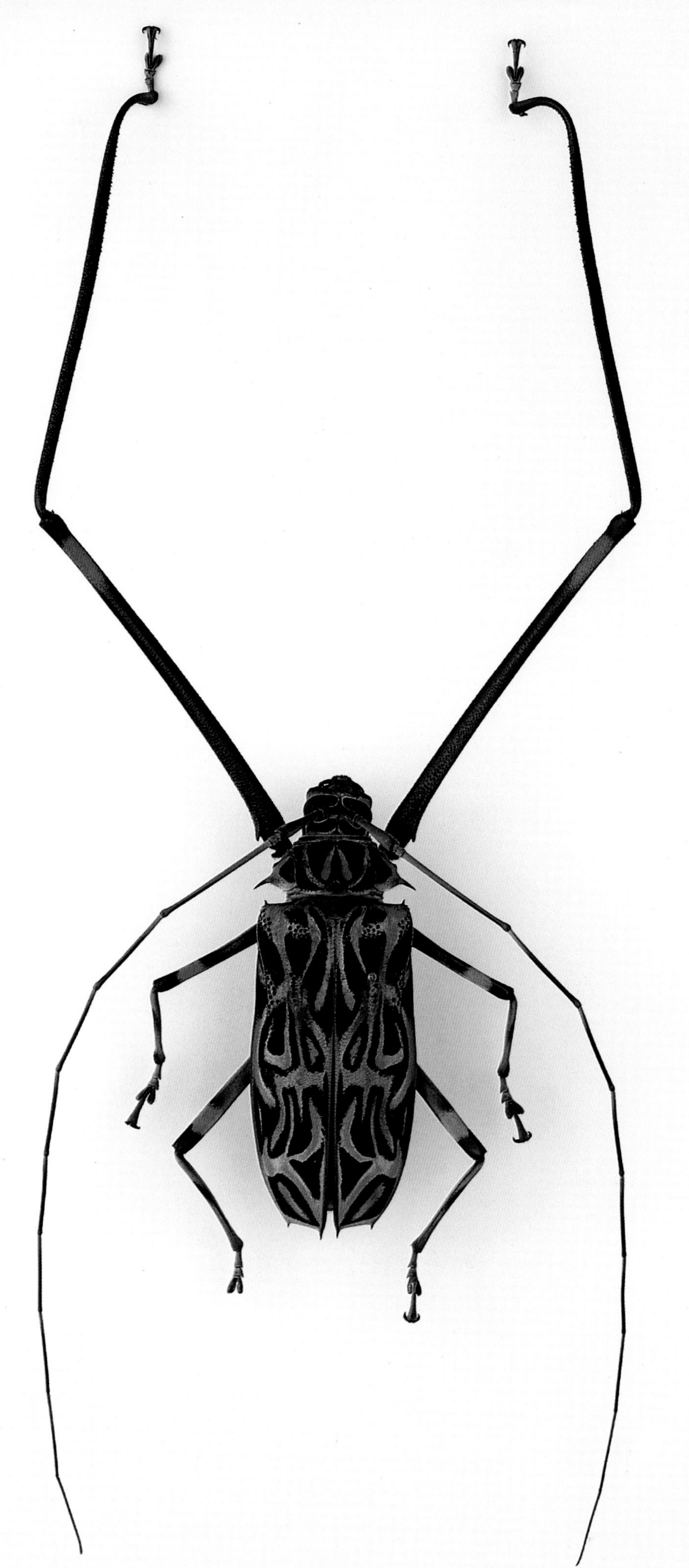

Opposite: *The scales of a butterfly wing, like this one belonging to an Indian nymphalid, may serve several purposes: to provide camouflage, to warn off predators from a toxic species (or its mimics) or to attract a mate. Wing scales, responsible for the insect's overall appearance, may even serve a strategic purpose and slough off—letting the butterfly escape—when wings accidentally touch a spiderweb.*

Pages 382–83: *Appearances—of a birdwing butterfly, say, like this one from the Solomon Islands—can mislead the taxonomist in many directions. Some of the papilionids have been named several times in the past because individuals which emerge at different times of year display distinctly dissimilar patterns. Consequently lepidopterists tend fertile females and rear their larvae under different conditions to compare the adult forms which result.*

Pages 384–85: *The uranid's wing scales are really transparent where iridescent colors appear. The cell structure of the clear scales acts like a defraction grid which reflects only certain wavelengths of colors. Early lepidopterists accepted color as the most important taxonomic character. The SEM now demonstrates that many scale arrangements can create the same color, perhaps for any of several adaptive reasons, while, conversely, similar wing-scale architectures can create very different colors. This implies that scale structure is a more useful, reliable taxonomic character than color, which probably is not true.*

earth remains a futuristic dream. The intrinsically fragmentary fossil record probably doesn't contain a single complete lineage even as far back as the Cambrian Era. But relationships remain important; the links among earth's tens of millions of living species must be deduced for reasons which are sound and understandable. As Duane Hope expresses it, "Valid taxonomy allows valid predictions" about an organism and its relatives. By contrast, invalid taxonomy creates chaos; imagine the merit of an explanation of flying fish mating behavior based on knowledge of birds and bees simply because they all fly. Of course, many animals that have developed traits in common—examples of convergent evolution—are harder to distinguish than exocoetidae, charadriae and colletidae. Taxonomists of different faunal groups have agreed on seemingly odd characteristics to distinguish various taxa: amphibians and dinosaurs by the pubic bone, mammals and reptiles by the number of jaw bones, bees and wasps by hairs that split or don't. When getting down to the nitty gritty, the taxonomist relies on his experience and familiarity with thousands of organisms to reach an informed opinion about which is related to what. Thus taxonomy's inherent problem, and its derision among some other scientists, derives from the fact that subjective judgment does play an important part. It makes no difference that physicians rely on insight and the great physicists on "the intuitive leap." (Perhaps because taxonomists can't awe the rest of us with miraculously cured patients or newly demonstrable natural laws to justify their work.) By definition, they contrive and fill dovecotes of categories— taxa (the plural of taxon)—only one of which has an absolute basis in nature. The rest of the construct is arbitrary.

The alpha taxonomist concentrates on the discovery and description of species that are new to science and, in naming them, assigns each to a genus which then presumably fits another cubby in the hierarchical structure. All this follows the convention established by the eighteenth-century Swedish naturalist Carl von Linné, whose name usually appears latinized (Carolus Linnaeus) in tune with his canon that every plant and animal receive a unique Latin binomial. By tradition alloyed with disciplinary muscle, the first binomial given any species in a scientific journal becomes its official handle—indeed that name by which all information

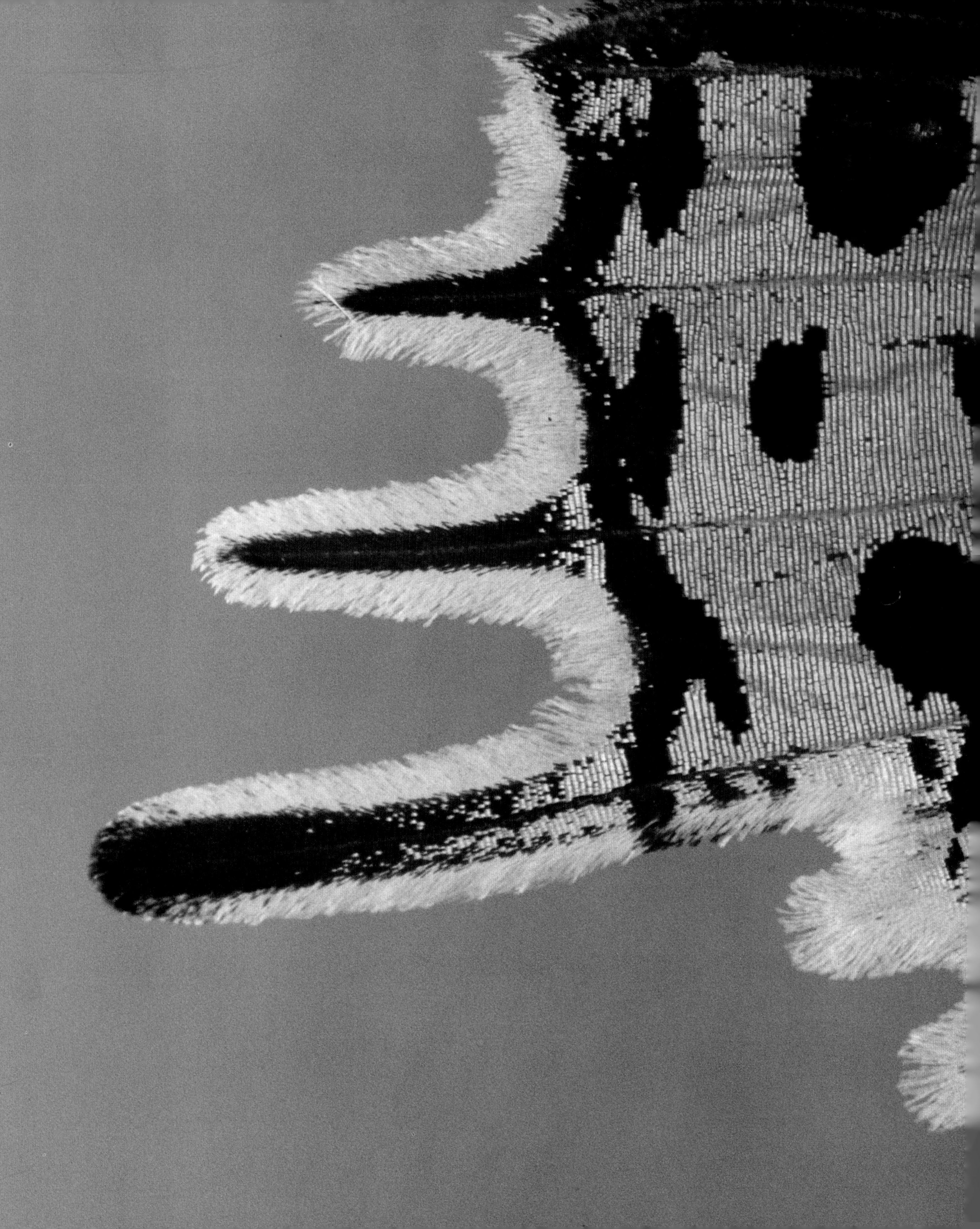

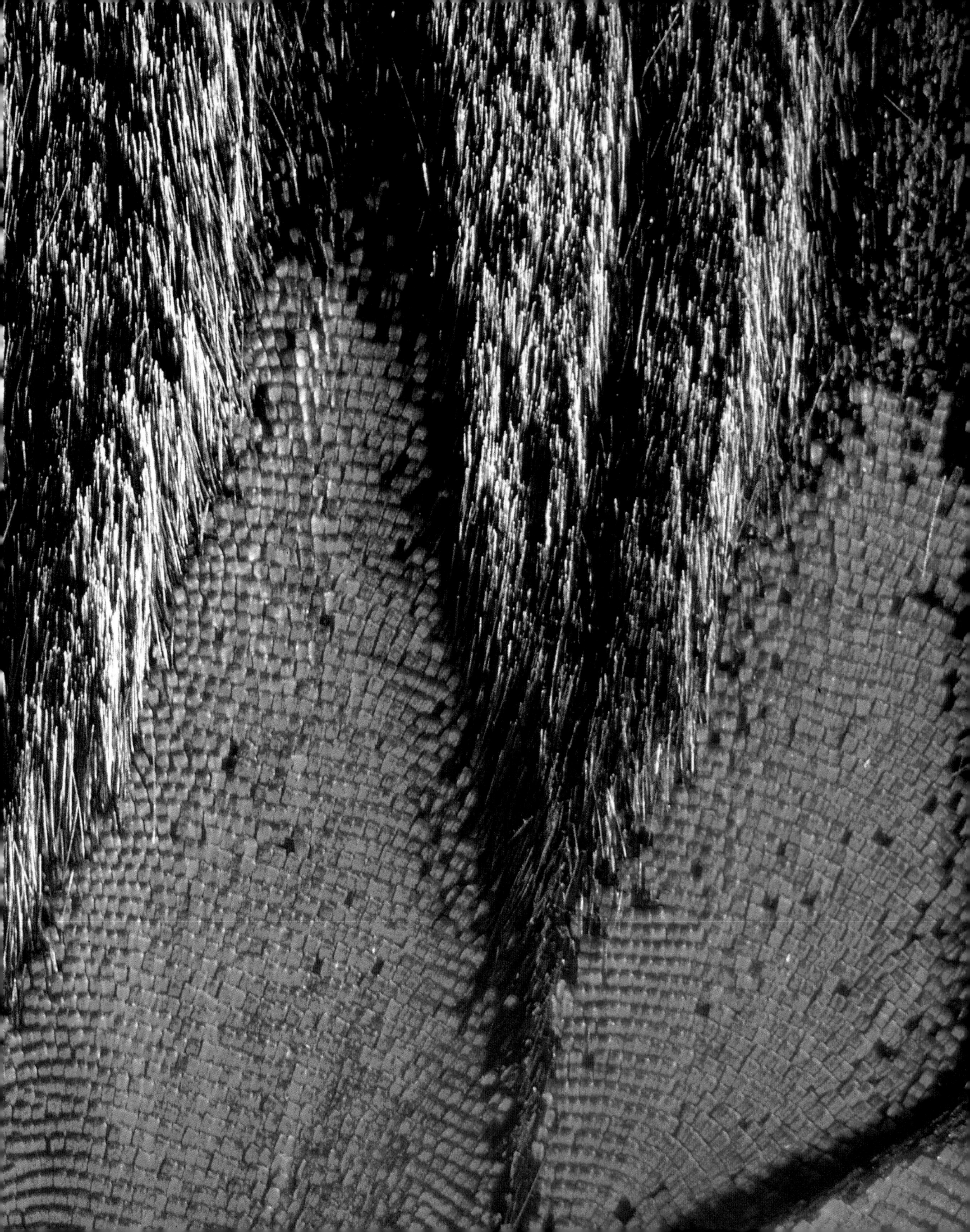

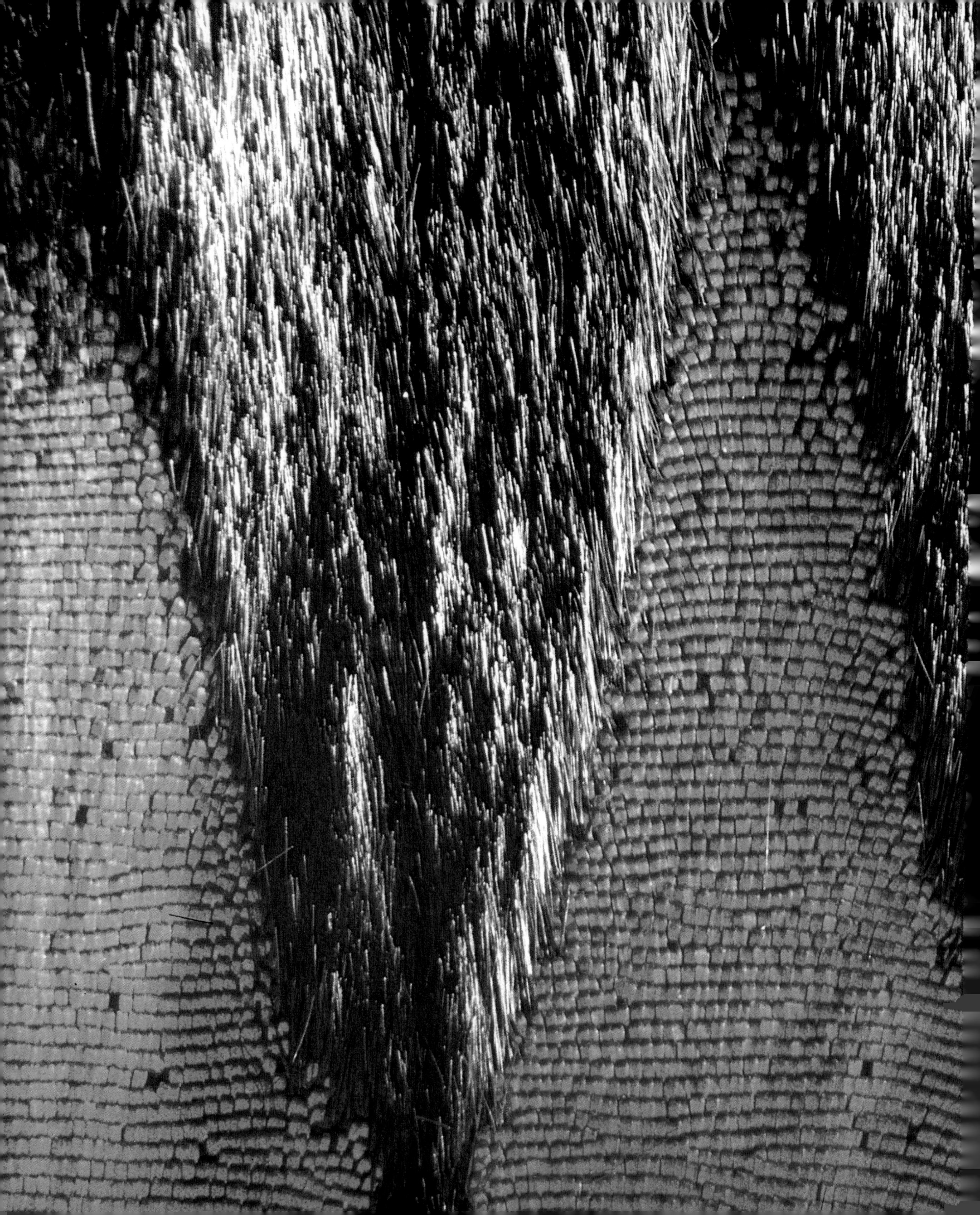

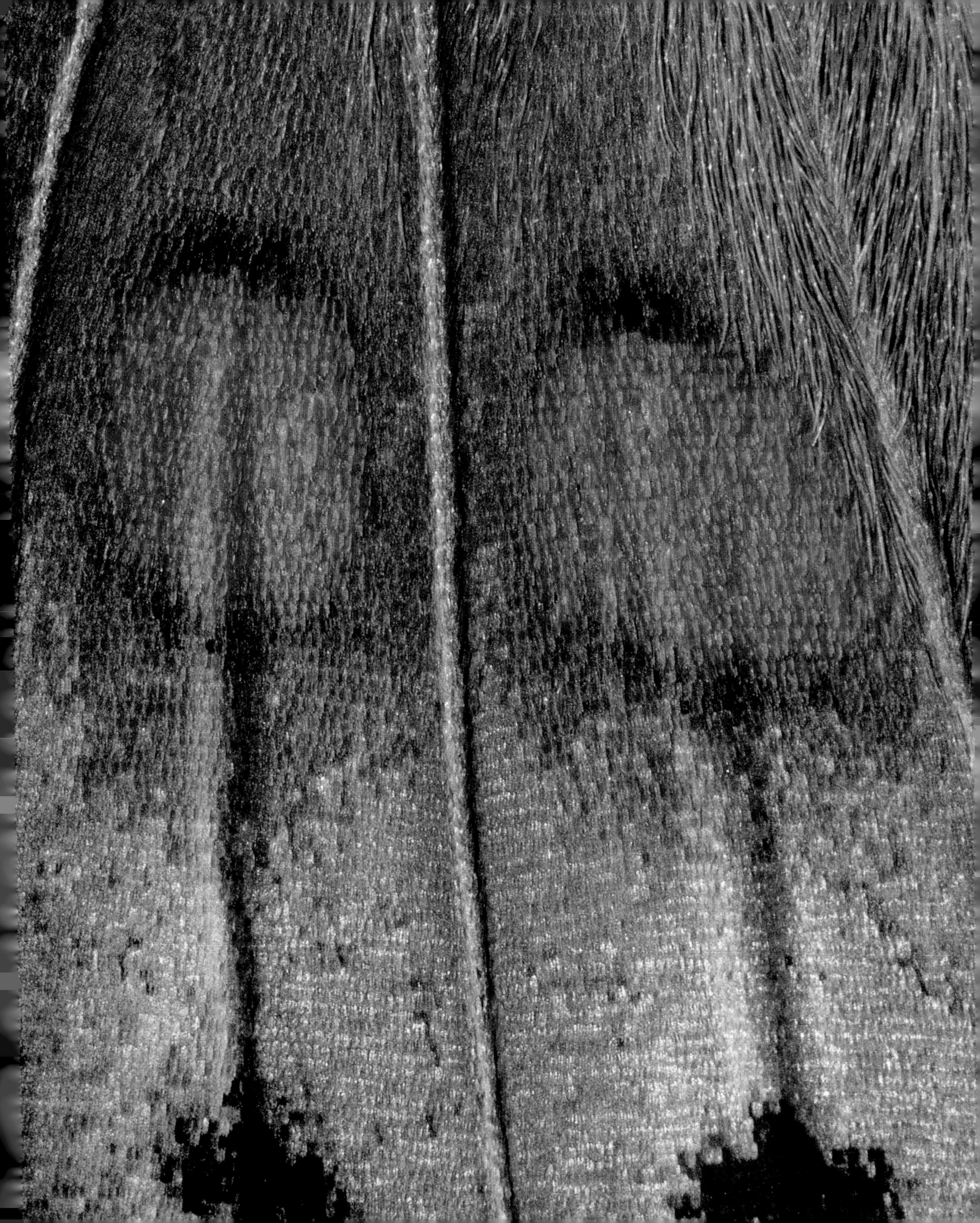

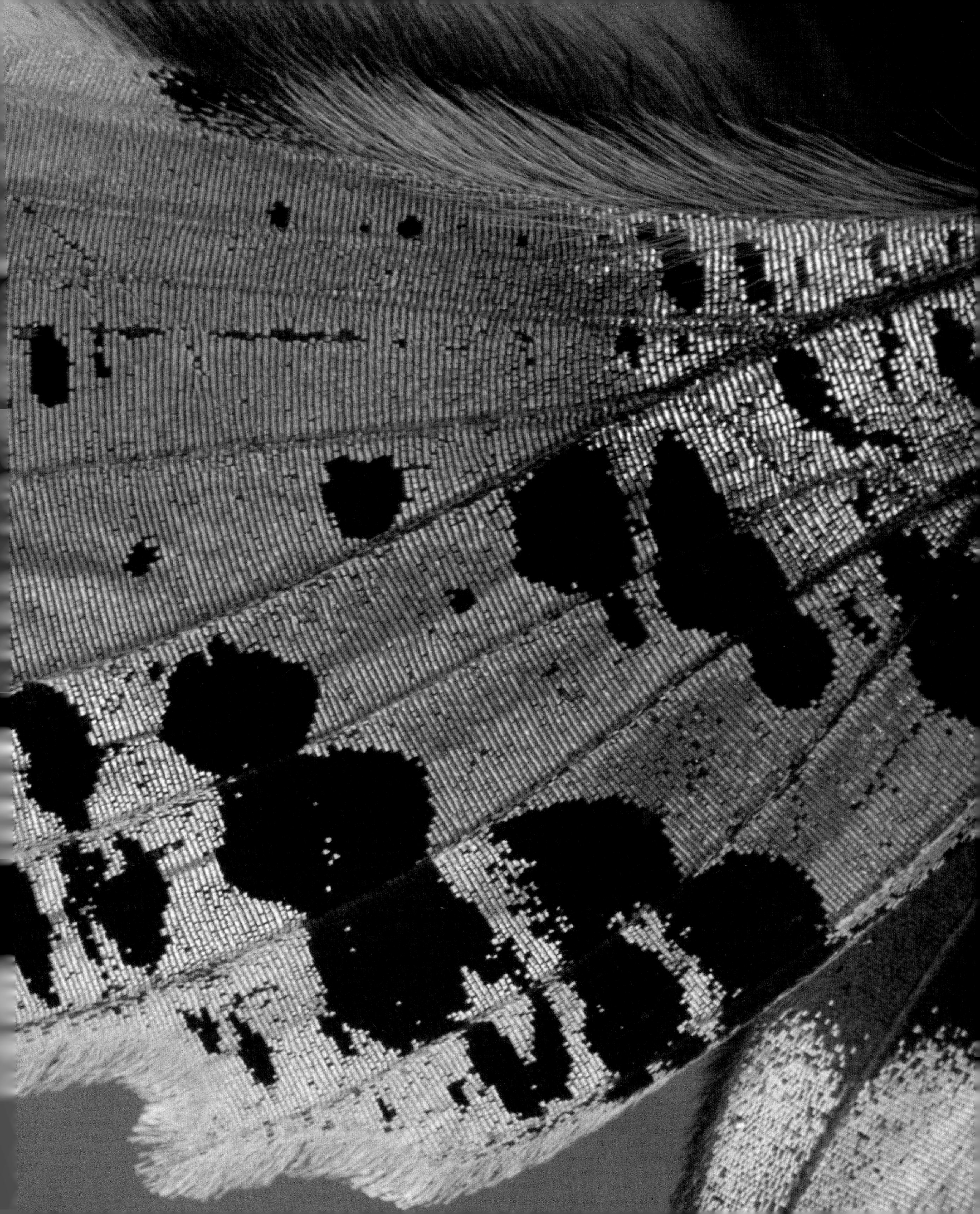

Opposite: *One curator on a two-month trip to Taiwan collected these moths and 50,000 other insects. They have been roughly sorted, mounted and fixed with tiny labels bearing the time and place of capture and the collector's name. A technician's next task will be to identify each one down to as low a taxonomic level as possible, though many may require a specialist's attention and description as a newly discovered species.*

Following pages: *Morpho butterflies seem to shimmer in a tray partly because they rest at different angles. Other distinctions between specimens reflect variation within a single species. All the Museum's biological departments maintain redundant collections—not just one example of each kind—because individuals of one species vary from each other in appearance. (Consider the variation of complexions among* Homo sapiens *or of the dot patterns on these butterfly wings.) One major task of modern taxonomy involves consolidating the many single species which were assigned multiple names by earlier investigators who lacked enough specimens to perceive the range of variation.*

about it can be compiled, filed and retrieved. When one organism gets named twice, or one name applied to two species by independent investigators, the Laws of Synonymy and Homonymy are adjudicated by an international commission on zoological or botanical nomenclature. Thus every organism introduced to science becomes known by only one name. If this system doesn't make anything simple, it works because it imposes some demonstrable order on the study of Nature's infinitely confusing variety.

The practice avoids the confusion inherent in common names. (In rural England an "urchin" is an insectivorous mammal, a hedgehog; in California it's a spiny sea creature. In Maine *Morone saxatilis* goes by the name "striper" or "striped bass," in Maryland by "rockfish.") The binomial, by indicating genus, also implies the animal's fixed place in the zoological scheme of things. As Carolus Linnaeus established the obligatory ranks, each species, i.e., *sapiens*, belongs to a genus (*Homo*), each genus to a family (Homonidae), each family to an order (Primatae), each order to a class (Mammalia), each class to a phylum (Vertebrata), each phylum to a kingdom (Animalia). But—and here's the snag—only the species has an absolute definition in nature. A species comprises those animals that can breed together in nature and produce fertile offspring. And even this gets a little dicey. The brown bear and polar bear can breed in captivity; but since their habits and habitats keep them apart in the wild, each is a species. Geographically distinct populations of a species, namely its subspecies, get an extra name, as in the western coyote, *Canis latrans latrans*, and the New England branch of the bunch, *C. latrans estis*. Where their ranges overlap they interbreed and confuse everybody.

Species aside, inherently arbitrary criteria delimit all other taxa. Members of a genus relate to each other in somewhat different degrees; they can only be said to be more closely related to each other than to members of another genus in the same family. Further, in each biological science, the different levels of hierarchy reflect different degrees of distinction and similarity. Returning to tires, as Akron's manufacturers have never agreed upon one set of standards for top-of-the-line goods versus the run-of-the-mill, so entomology and herpetology have different

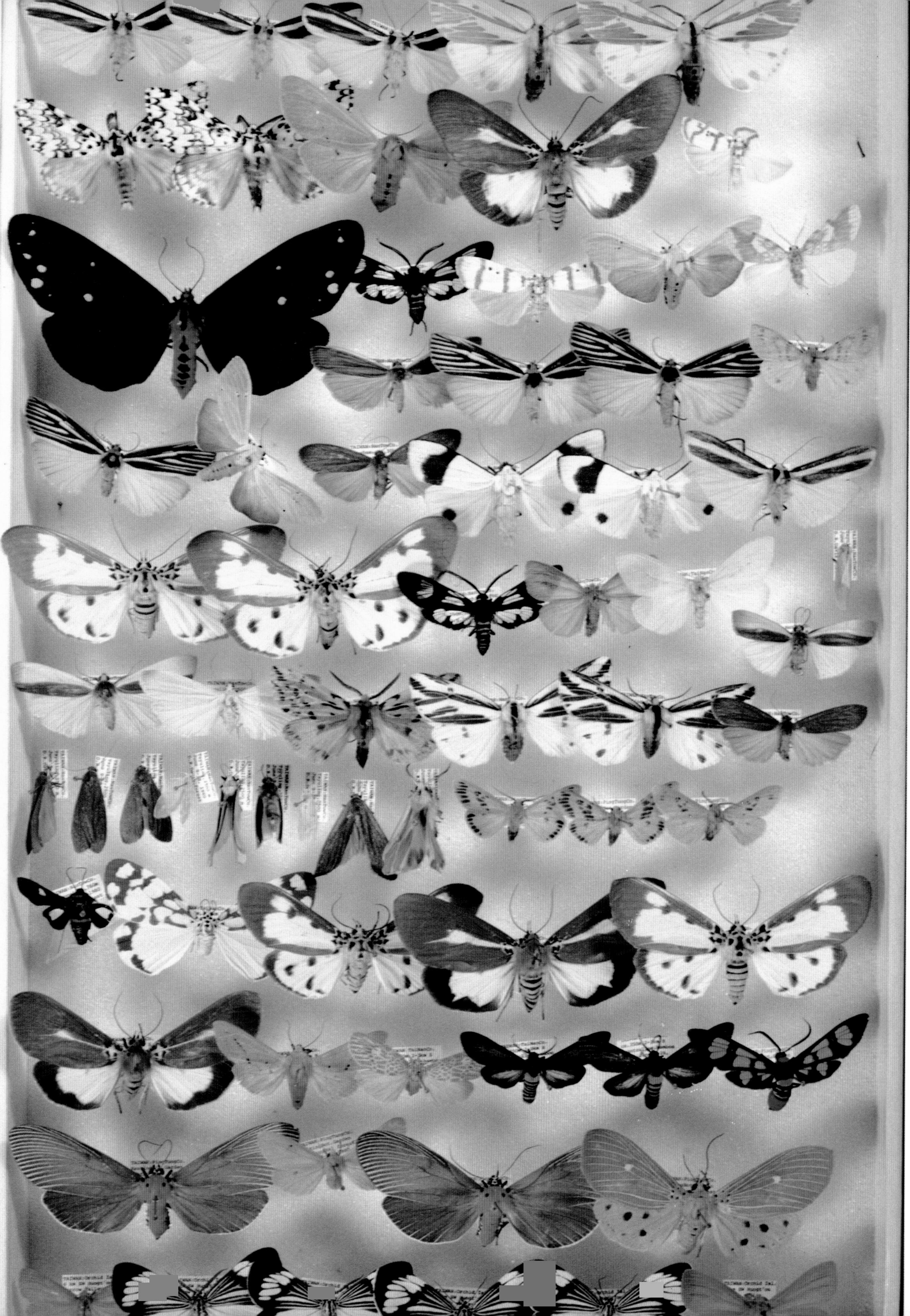

Opposite: *The front and back of a member of the Pieridae family reveal splashily colorful upper wing surfaces which are visible when the insect flies. When it settles and folds its wings together, the rearmost pair swings up slightly, covering the bright region of the front pair. Only the cryptic, camouflaging design remains visible.*

Following pages: *This tropical American "silk moth" hatched from a cocoon but—despite* Rothschildia*'s common name—not one that produces lustrous thread. Some ancestral arthropod larvae probably contrived proto-silk to fashion shelters out of leaves. As insects radiated, the kinds of silks and cocoons they made diversified too. Ancient China domesticated a caterpillar which produced useful cocoon material and the first Western taxonomists lumped every remotely similar insect in the "silk moth" group, an aggregation that proved too vague as the science matured. The common idiom persisted though only one species makes a silk strand that can be unraveled in large quantities. The true silk moth,* Bombyx mori, *was selectively bred—for silk production at the expense of all else—over so many generations, let alone centuries, that it lost the power of flight and no longer lives in the Asiatic wild.*

criteria for what distinguishes families from orders, et al. Finally, as biological classification has grown, Linnaeus' obligatory ranks proved inadequate, and so intermediate classifications evolved, such as subkingdom, superclass, infraclass, cohort, tribe and so on.

If all this appears a hopeless muddle, keep in mind the unidealized, scientifically practical goal of taxonomy, systematics and classification: to contrive a workable system for cataloguing all the organisms in the world—a necessarily synthetic system designed to reflect natural relationships as closely as possible. Earth's biota grew like Topsy over a few billion years without anticipating one species' eventual bent for intellectual order and analysis. Early organisms gave rise to specially adapted and specified groups of offspring, which in turn radiated further. The challenge facing taxonomists of every alphabetic category is to discover and describe different degrees of kinship.

Just as librarians can debate the relative merits of the Dewey Decimal and Library of Congress catalogue systems among themselves, so taxonomists disagree about the established classification system(s). Three schools currently compete: Numerical Phenetics, which tries to weigh all physical characters with arithmetic objectivity; Cladistics, from the Greek word for "branch," which graphically depicts relative genealogy; Evolutionary Classification, which attempts to combine descent and divergence and in doing so frankly depends on some informed subjectivity. Each has some strengths, some weaknesses and an imperfect name, which just confuses things all the more. Suffice it that no less a personage than the Alexander Agassiz Professor Emeritus at Harvard's Museum of Comparative Zoology recently suggested a compromise: Let "the best components of the three methods be used. By such a synthetic approach, classifications can be constructed that are equally suited as the basis of generalizations and as an index to information storage and retrieval systems." That last sentence almost says it all. In this imperfect world of imperfect science, taxonomy has finite and useful goals: to enable the categorization of organisms into groups so that generalizations can be made about them; then to efficiently store and easily retrieve the information. Any outsider who tries to wager on the absolute superiority of one school or another loses sight of the

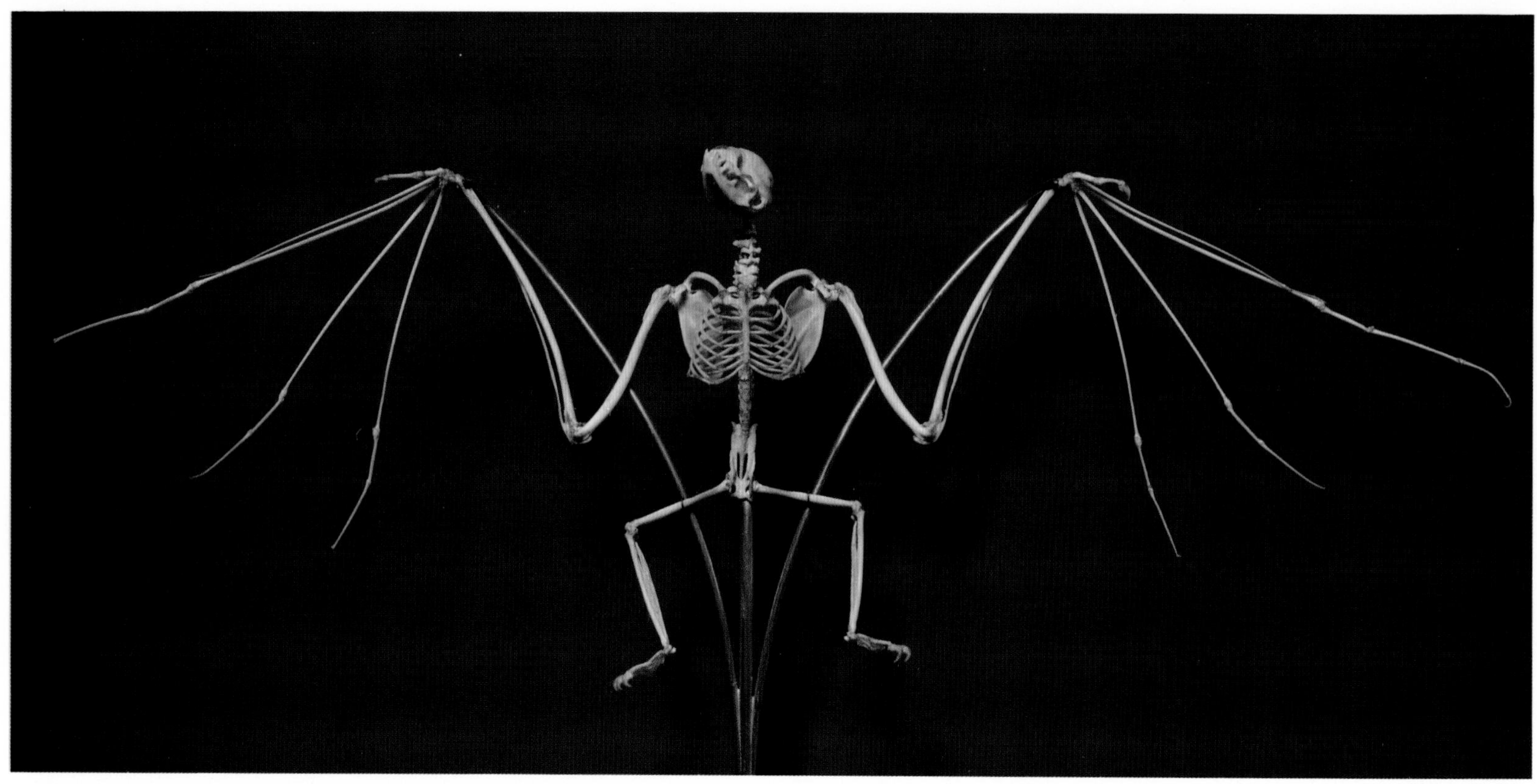

Above: *The little vampire bat* (Desmodus rotundus) *with its ten-inch wingspan lives on the blood of other mammals only in tropical America.*

Opposite: *The flying foxes of Southeast Asia (e.g.,* Pteropus samoensis*) boast wingspans of 5 or 6 feet and live exclusively on fruit.*

fundamental commonality among all genuine taxonomists regardless of nominal affiliation. As one Museum botanist puts it, "the scholarship is more important than the terminology."

For the moment, according to another authority, "There is no single theory of classification that a biologist is obliged to accept and use. . . . There is no generally accepted theory of classification." Which may be what makes that taxophobic geologist so mad. But neither is there a single theory of verbal communication, let alone a single dominant language among the world's people. Yet we manage to talk and listen. More important than the internecine disputes, taxonomy has important uses. As Richard E. Blackwelder writes: "Taxonomy is an observational science, part of the age-old field of natural history. It is concerned with knowing animals as individuals, kinds and groups of kinds. It is historical in large part, studying the work of the past, adding to it, and reorganizing it to yield new insights. Although the goals of the individual taxonomists may differ, the science itself has a triple goal: to systematize the knowledge of animals [and plants and protists], to add to that knowledge by observation and analysis, and to make this organized knowledge available to all biologists."

The arguments have become so shrill and esoteric that even some taxonomists plug their ears. Fenner Chace, who came to the Museum nearly 40 years ago, remembers when "research was a dirty word around here and curators tended the collections." That required identification and

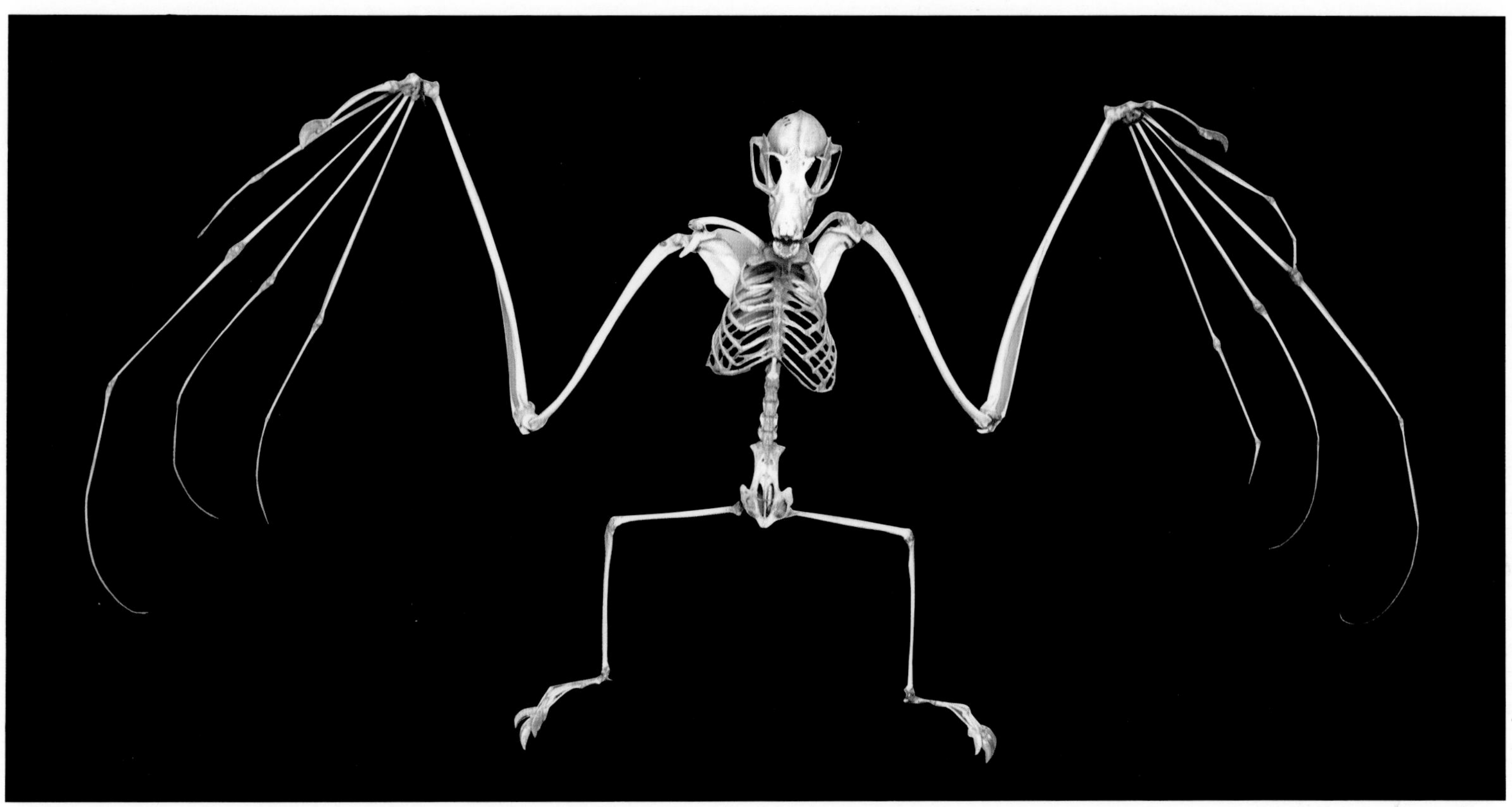

naming of new species but little higher systematics. He thinks that science was better off then, because conflicting schools of theoretical thought have distracted biologists from the more important work. Who started the trouble? "Darwin," Dr. Chace replies instantly. Before he became a household word, scientists busily catalogued everything they could find Monday through Saturday and joined their peers in dozing through sermons on the Sabbath. "They accepted the fact of endlessly varied species as a manifestation of divine bounty," Dr. Chace believes, until *The Origin of Species* rent the intelligentsia asunder and theological harangue clouded scientific colloquy. The describer of evolution would have been better advised to stick to taxonomy, i.e., his classic and still instructive study of the world's barnacles, rather than shift the investigative emphasis from "What?" to "How?" The study of animal kinds has been on the wane ever since, Chace says. Very possibly taxonomy will never again germinate anything as significant to man's thinking as the theory of evolution, "but it still has a tremendous and potentially important job to do." (Chace himself now studies crustaceans collected by the *Albatross*, which embarked on a Pacific cruise in 1907, and finds a pleasant irony in knowing they still lived during the year of his birth. If his descriptions of them hold up so long, or as long as Darwin's monograph has remained valid, he'll be content.)

Dr. Lewis Thomas, the celebrated physician and savant, believes this century's landmark achievement is the recognition of human ignorance. To put it another way, man is just coming to see the limited extent of his

Opposite: *A clutch of tiger coweries displays the range of color and pattern variation in one mollusk species. The genus* Cypraea, *to which* C. tigric *belongs, was named for the goddess of beauty, Venus, with a latinate term for Cypriot after the island of her mythical birth.*

knowledge. This is especially true in taxonomy and systematics. As for the identification of species, only a fraction have even been named; as for their relationships, many of the old arrangements require revision. All things considered, the tree of life has only been drawn in the broadest strokes with patches of detail. A simple limb, like the one representing echinoderms, appears relatively clear up to the point of joining the trunk. But the kinship of algae is so obscure, says botanist James Norris, that specimens in the Herbarium are organized by pigment (the traditional, fundamental and still valid primary breakdown). Then here in the National Herbarium at least, the species of each color are arranged alphabetically since few phycologists agree on how they're all related.

While scientific debate (and sometimes deafening unscientific noise) continues about just how and when *Homo sapiens* split off from the other primates, mammalian phylogenetics seems relatively clear. Yet even this realm still contains surprising secrets. Mammals curator Richard W. Thorington, Jr., only recently discovered that an antique squirrel which lived at least 35 million years ago has a nearly identical descendent, a "living fossil" relative, in the fox squirrel, the common gray squirrel's closest cousin. Thus the mammals' descent and their degrees of divergence have only been partly defined at best, though probably 99 percent of Class Mammalia has been identified and named. (What does the remaining 1 percent represent? A few tropical shrews, no doubt; some particularly cryptic rodents, members of the order comprising fully half of all mammalian species; several bats, the order containing half of all the rest. As for Big Foot, a.k.a. Yeti and the Abominable Snowman, Dr. Thorington, who studies the reported sightings around the world, has his doubts.) The situation is markedly less complete, therefore worse, in almost every other phylum. In those major groups considerable alpha taxonomy remains to be done from scratch, and the higher systematics must be revised.

Clyde Roper, who studies cephalopods around the world, admits his discipline suffers from the surplus enthusiasm of its pioneers. They tended to treat every new giant squid they encountered as a new species simply because its tentacles might be a few yards longer than the last one's.

Opposite: *A random selection of insects arranged for display in the Evolution Hall shows a fraction of these arthropods' diversity. Many experts estimate between 3 and 10 million members of the insect class live today. A curator who has found new habitats in rainforest canopies puts the total number at 30 million species.*

Conversely, what fishermen in one country consider to be just one vari-colored squid often turns out to be several species. This can ruin efforts to expand a local fishery until the animals get sorted out, says Dr. Roper, who frequently advises governments on these matters.

Even after fully 1 million insects have been described, a former Entomology Department chairman believes that barely one out of ten living insects are known, while another authority believes the ratio may be closer to 1 in 30. (Collecting in the Amazon basin, Terry Erwin spreads sheets on the rainforest floor, fires an insecticide bomb loaded with lethal chrysanthemum extract into the 100-foot-high canopy and gathers hundreds of species never seen before. Previous generations of entomologists never figured out a way to collect at these heights.)

In part, fairly simple inventions have led to new advances. Earlier entomologists had to be content with collecting whatever they happened to see, reach and catch. Now they visit promising habitats equipped with battery-powered ultraviolet lamps mounted atop baffles which tumble flying insects into unattended traps. While the collector camps out in relative comfort, the trap works all night unattended. (Many insects may be attracted to light because they hatch in dark cocoons with only one exit; the light of their first day shows them the way out, and light's attraction remains lifelong. Some night fliers may perish in a candle's flame because the light immobilizes the wing on the same side that the light appears, another function of their primitive neurology.) The study of bats has been revolutionized through the unlikely agency of General MacArthur's occupation of Japan. For 300 years, says curator Charles O. Handley, Jr., the Japanese used invisible mist nets made of silk to catch songbirds for the table. Western chiropterologists discovered the secret after the war and had the nets knotted out of synthetic fibers. Then, for the first time, they could get samples of entire bat populations, instead of only studying the species that roost in colonies. As a result, in some tropical regions as many new species of bats were discovered as had been previously known. (Bats constitute the most numerous mammal group in the tropics.) The impact rivals that of the mousetrap's invention on rodent studies in the 1880s. Until cheap, portable traps could be set out, naturalists only knew those

mice that cats dragged in and tended to identify each newly encountered color pattern as representing a new species. Once entire populations could be sampled, they could see the degree of variation within each species.

Tools aside, zoologists of every discipline have made important strides by just working more carefully and collecting more comprehensively. For instance, 20 years ago 175 species of coral were known from the western Atlantic. Since then more systematic dredging has produced almost as many more new species. There is a study of one oak tree in California. A fastidious investigator found 17 insects on the tree; only three were known—that is, they had scientific names. In botany, 250,000 living species of vascular plants have been named, but an estimated 100,000 remain unknown to science. In addition, there are 100,000 known fungi and probably a like number of unknowns.

The practical implications of these vast gaps in biological data—voids that alpha taxonomists mean to fill—are frightening or absurd, depending on one's viewpoint. It goes without saying that our blissful dependence on 30-odd species of plants worldwide for food is both dangerous and shortsighted. (Edward Ayensu, director of the Endangered Flora Project, says that man has only utilized about 3,000 species throughout history and today gets fully half his food from three: wheat, rice and corn.) Innumerable other forms of vegetation have potential uses, especially as foods and drugs. Similarly, countless insects have the potential for becoming agricultural allies if their prey becomes pestiferous in a region, just as a like number—some of them the same insects—might become uncontrollable pests if their predators disappear. To deal effectively with these problems and promises requires detailed knowledge, both of the critters and the plants involved. But to know them is first to name them. Thus one of the primary concerns of the Botany, Invertebrate Zoology and Entomology departments in particular involves taxonomy pure and simple. Much of the work crowding their endless agendas occurs at the elementary level of identifying species which are new to science though they may be ancient on earth and represented in unsorted collections for decades.

The upshot of all this intensifying and improving work: taxonomies and systematic arrangements frequently do get revised, as the geologist

Opposite: *Like a jewelbox in a quarry, the Botany Department's greenhouse glows in the Museum's East Court. This small hothouse lets plant taxonomists keep some living material close at hand to study in various growth stages what cannot be studied with dried specimens in the herbarium. Its presence highlights an aspect of taxonomy too often lost on outsiders: While this science concentrates on the description and classification of organisms both extant and extinct, many of its most dramatic applications involve living plants which play crucial roles in agriculture and ecology.*

Opposite: *The fern* Alsophila sinuata *bears distinctive berry-like clusters of spore-bearing organs called sori. These structures prove more useful than the fronds in classifying this Sri Lanka species and discovering its relationships to other ferns.*

complained. But not for purely capricious reasons. Revisions occur for the same reasons that new, "completely revised" editions of geology texts appear: because they contain new information and reflect more complete understanding of the subject. New data, more sophisticated investigative techniques, more careful study or a more complete inventory of specimens can render an old taxonomy obsolete. The discovery of new evidence—a fossil or living species representing an unexpected missing link—can turn an accepted phylogeny into dated science fiction. Drafted with the best intentions and according to the established criteria of the day, it has been overtaken by scientific events.

Inevitabilities such as these pose problems, especially for outsiders: taxonomy often seems too dynamic for its own good. Ordinary folks asking questions of scientists want straight answers guaranteed to have shelf-lives as long as books bound in buckram. But because of the progressive nature of science, many answers seem as short-lived as quarks. In fact, scientific truth is transient, alas. Every hypothesis, when repeatedly proven and elevated to the capitalized status of Theory and then Law, awaits substantial revision, particular restriction or even repeal in the light of new knowledge. (When I first studied science the world happily accepted the existence of 92 elements that were regarded as different combinations of three kinds of particles. Now there are 103 known elements and the atom, once authoritatively considered the smallest indivisible particle in nature, comprises whole new families of constituent parts. Even matter itself no longer stands alone but has acquired the nebulous partner of antimatter. No doubt it existed before, as intangibly as today; we just didn't know about it.) Newton's Laws of Motion served well enough to govern visible mechanics for two centuries; Einstein's Theory not only illuminated the atom and its massive energy but explained why orbiting planets and expanding galaxies don't act like apples falling on sleeping theorists' heads. *Sic transit leges naturae.* So long as our knowledge remains incomplete—and that it will remains a singular certainty—the laws of nature as codified by man will be subject to amendment. So let it be with taxonomy, a discipline that addresses many challenging questions and answers them usefully pro tem.

Opposite: *One of some 4 million specimens in the National Herbarium, this Sumatran "bird's nest" fern frond has been sewn to the paper to keep it attached without breaking.*

Some alpha taxonomists believe that they practice "the world's oldest profession." After all, the first task Adam turned his hand to was nomenclature. Within 13 verses of his own creation in the second chapter of Genesis, "Adam gave names to all cattle and to the fowl of the air and to every beast of the field," according to Judeo-Christian tradition and the King James Version. Alas, he hadn't catalogued all the plants, marine animals and insects before Eve beguiled him with the apple and they lost their tenure in Eden.

While many practitioners collect fresh material in the field every chance they get, new species still often appear in old collections. Entomologists and botanists in particular aren't content to study only the material on hand until it's all been identified; they can't be. The reason: species that still survive in the wild today may be gone tomorrow. Entire habitats, especially in developing and tropical countries, are being razed in the name of progress. Rainforests are being axed for timber and fuel or to make space for cattle ranches, mining operations, suburbs. Consequently, plants and insects must be collected now; they may be extinctly uncollectible in only a few years. Curators and some technicians alike gather new material at their own expense, knowing there will eventually be time to assess it all while the time runs out for gathering uniquely adapted species from unique microhabitats and surviving relics of ancient ages.

Meanwhile the intramural work proceeds. In entomology technicians sort material gathered decades ago, labeled only as to when and where each specimen was collected. Tray after tray of insects then goes to specialists for detailed study and identification. By tradition and explicit agreement, the Museum's insect people divide the work in a pragmatic fashion. About three dozen specialists, most of them armed with doctorates, focus on the insect groups whose members have particular relevance to agriculture (for good or ill). These people constitute the staff of the Agriculture Department's resident Systematic Entomology Laboratory, which has worked hand in glove with the Museum since it first started studying insects. The Museum's department itself has only a dozen entomologists and they handle all the other groups except mosquitos, the monopoly of the Medical Entomology Project.

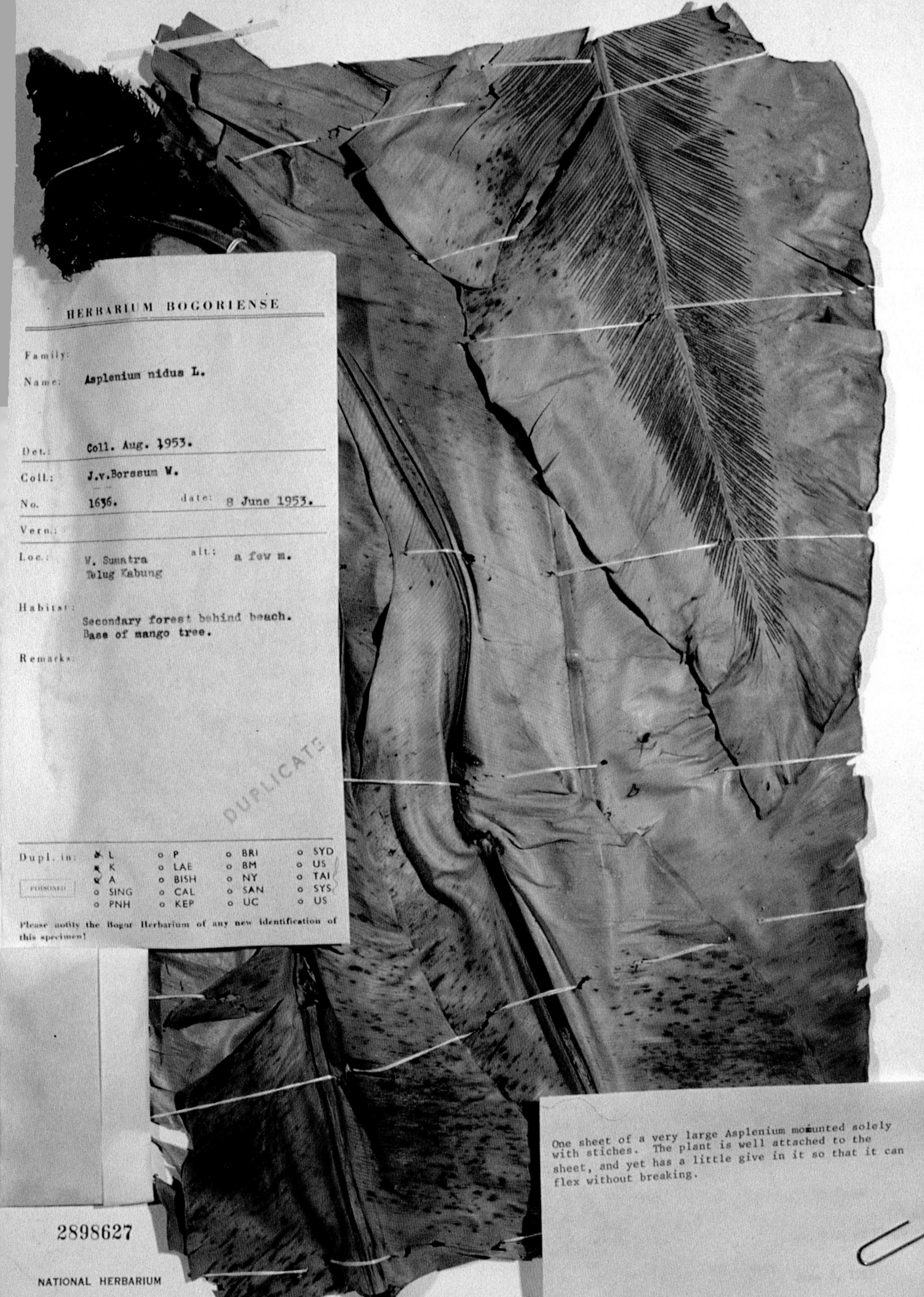

One sheet of a very large Asplenium mounted solely with stiches. The plant is well attached to the sheet, and yet has a little give in it so that it can flex without breaking.

Opposite: *Another fern from the Herbarium, heart of the Botany Department, has been attached with strips of gummed linen. The frond is folded so that both its upper and lower sides may be examined by taxonomists.*

The question arises, why don't the Agriculture Department investigators limit themselves further and only study agriculturally significant species within the groups of weevils, moths, flies and aphids they handle? Because just as any scientific truth is transient and depends on the surrounding state of knowledge, so any kind of economic or medical significance depends on purely temporary conditions. Start studying half a dozen leaf miners which happen to be ruining an apple crop and by the time a pesticide has been applied, six other species, perhaps immune through some shared genetic trait, will start getting fat and multitudinous on Golden Delicious leaves while their natural controls succumb to the chemical sprays. This shortsighted answer suggests the real issue: there's no way to purposefully study a few temporarily pestiferous "true bugs" without understanding their place among the other 20,000 species in the Hemipteran Order. (The sheer numbers might be enough to suggest the difficulties, but there are more. Thomas J. Henry, Agriculture's accurately described "bug man," recently demonstrated that two presumed species were really one; the original female had been misidentified.)

The greater purposes of taxonomy aside, the discipline frequently has immediately important applications. Penicillin could not have been used as a drug until pharmacologists learned from botanists how to identify the single mold. Public health officers in Europe wasted fortunes and years trying to wipe out a nation's mosquitos—until entomologists discovered that one species of five seemingly identical ones was responsible for malaria. (The single culprit could only be identified by its eggs.) Some years ago a New World cactus was imported to Australia as an ornamental plant; it spread over 50 million acres in 20 years until a search of the scientific literature turned up two dozen parasites. One of them, described 80 years earlier by an Argentinian, was imported and virtually wiped out the cactus. In Florida, a power company's underwater electrical cables were repeatedly damaged by a marine mollusk; the species turned out to be unknown, but a systematist identified its genus as one that lived atop the substrate. When the power company began burying its cables in the mud, the short-circuits stopped. More recently, a food packager has come hat in hand to the Museum. The firm, famous for its fried clams, is concerned

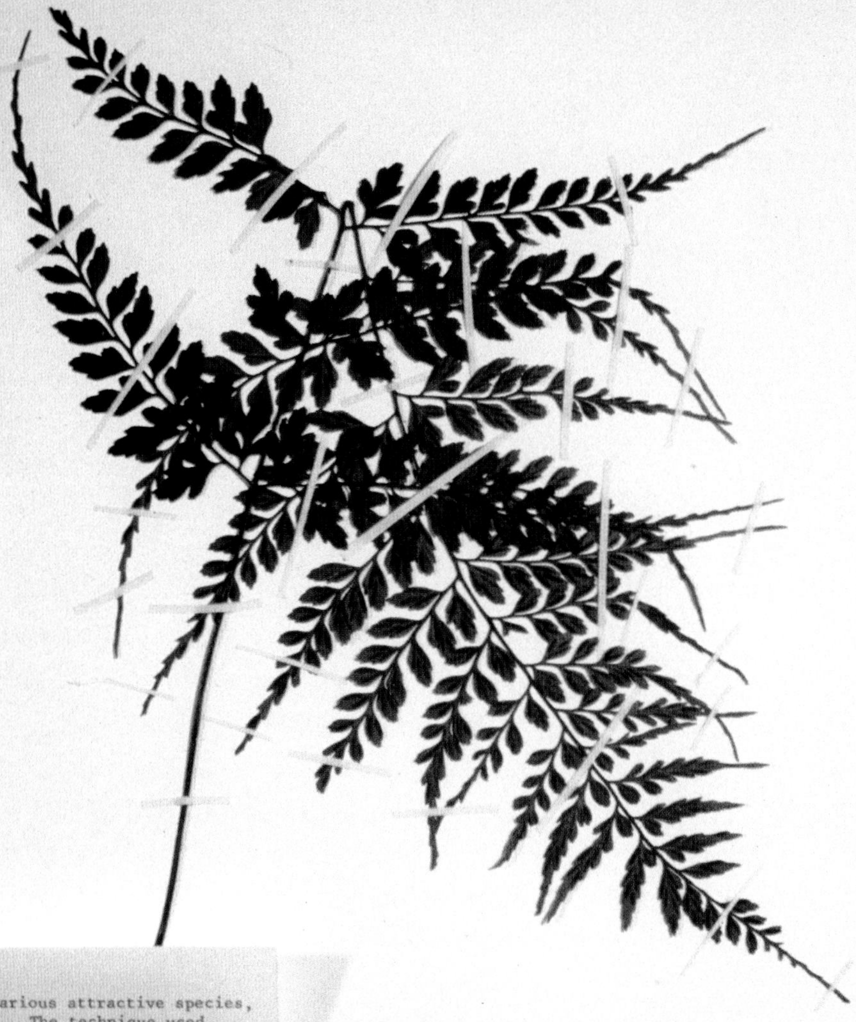

Three sheets of Asplenium, various attractive species, recently mounted with strips. The technique used here is not too good; the strips are long, and likely to detach over time. (Compare the shorter strips used to mount Lellinger 355 (Thelypteris dentata)).

FLORA OF NEW IRELAND

Botanical Collections of the Division of Botany, Department of Forests, Lae.
LAE 65276 J. R. Croft 4/8/74

Locality: East side of Lake Myola 2. West slopes of Mt. Kenive (Nisbet) Subdist Kokoda, district Northern. Altitude 2,000 m.

Lat.: 9 10 s **Long.**: 147 45 E
Habitat: Submontane rainforest

Annot. Low epiphyte. Fronds semigloss dark green above, mid green below. Sori brown, rhizome with dark brown scales.

UNITED STATES

2924507

NATIONAL HERBARIUM

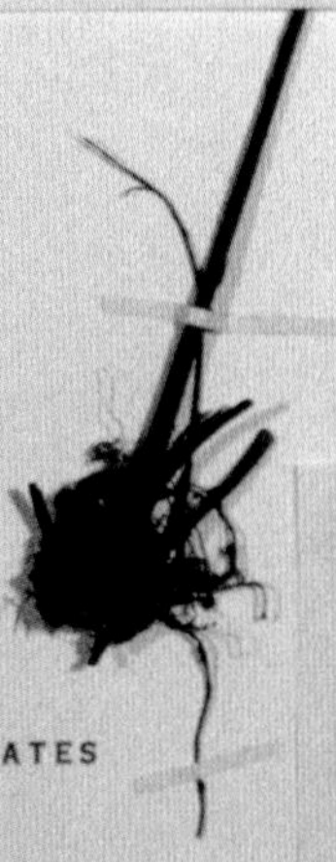

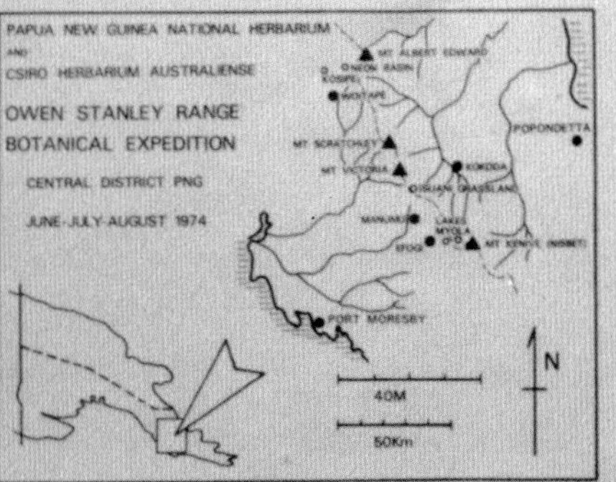

Fam.: Aspleniaceae

Name: Asplenium

Dupl. sent to: L. Bri. Canb. A. K. Bog. Sing. Syd. UH. PNH. US. Bish.

EDIN

about dwindling supplies of the species it has always used and hopes to find a palatable, plentiful substitute.

Lest one infer that this discipline involves only matters of arcane scientific interest, particular economic impact and endless tedium, consider the goings-on that occur when a systematist or taxonomist gets down to choosing a name. The opportunity arises to honor a colleague or mentor. Former secretary Alexander Wetmore has namesakes in 18 birds, three mammals, nine insects, 15 marine animals and a glacier. Entomologist Paul Spangler, who has named almost four score new species himself, knows of more than 40 named for him including a genus of whirligig beetles dizzily known as *Spanglerogyrus*. When Spangler encountered a new species in the South American beetle genus named with the local word for waterfall, he recognized its rough lip in its specific name. Thus was christened *Ytu brutus*. Crustacea curator Raymond B. Manning discovered a crab whose restricted carapace constantly exposes its copulatory organ. A Greek dictionary provided scientifically acceptable terminology, so the animal was named *Enchantor modestus*, or "modest flasher." With the infinite patience that taxonomists are noted for, senior technician Maureen Downey awaits the appearance of a member of the clan she knows better than anyone else in America. It will be named *Asterias mico mica*, for "twinkle twinkle little star."

Tradition requires that a name have some scientific relevance, but if taxonomists are remarkably patient, they sometimes overtax their imaginations or simply run out of nomenclatural steam. Whole genera have been given anagrams of their discoverer's wife's name, or anagrams of their habitats, i.e., Africa or Carolina. One organism was named *Golfingia*, "in honor of golf." To help colleagues come up with better names, the late and legendary paleontologist Roland Brown wrote a curious book, *Composition of Scientific Words*, and published it at his own expense (though he was, among other things, one of the Museum's famous misers). In this fat lexicon of terms in several languages which might be applied in nomenclature, he brightly argued against unspeakable complexity, such as the names chosen for two amphipods found in Lake Baikal: *Leucophthalmoechinogammarus crassus* and *Siemienkiewicziechinogammarus siemien-*

kiewicz. Brown also wisely objected to boringly redundant "simplicity, simplicity, simplicity" (to borrow Thoreau's emphatic replication) and took issue with designating prime subspecies as *Bison bison bison, Cardinalis cardinalis cardinalis* and *Rattus rattus rattus*. He concluded: "Nomenclature need not compete with the Hallelujah Chorus."

"It's fun," says one taxonomist of systematics work in general. Fun not only in terms of inventing names, which is only icing on the cake after all, but intrinsic and sober fun in terms of exercising deep curiosity and employing extraordinary care to meet a challenge: to fit together pieces of a puzzle that wasn't designed to have a solution. Finding one can demand remarkable ingenuity and a dazzling variety of skills, as W. Ronald Heyer displayed in sorting out a group of South American frogs.

Since mating calls play a crucial role in frog speciation, Dr. Heyer first studied the frog hearing apparatus and learned that it functions very differently from any mammal's. "Frogs act more like crystal sets." They only hear very narrow frequency bands, ones that are genetically tuned, but each frog has two "crystals." One of them, probably tuned through the skeleton, monitors low-frequency sounds such as those made by the tread of a passing predator. The other, tuned through the eardrum-like tympanum, only registers sounds in a high-frequency range. But since several species hear (and employ) the same pitch, something else must be involved or individuals of one species would go crazy responding to the calls of an alien species. The other factor proved to be the temporal patterns, or beat. Among very closely related species—a taxon below the genus level that he calls "species group"—timing means everything. Each listener's ear, constantly set to one frequency, registers only "on" and "off." Thus these frogs use something like Morse code when sending and receiving vital messages of their own kind and ignoring their congeneric cousins.

By analyzing the different codes, Heyer could differentiate 21 intimately related species, but couldn't quite unravel their precise relationships. This required biochemistry, the calibration of "immunological distance units" and a collaborator because as organisms evolve they change in both obvious and subtle ways, such as the amino acid make-up of various proteins. The sequence of more than 500 amino acids linked together in the albumin molecule changes at an average rate of one every million years. A biochemist developed an assay technique so precise she could spot the displacement of one amino acid in a chain of 580. With her help, Heyer worked out the albumin compositions of closely related frogs, then extrapolated back through time to estimate when two nearly identical modern species had the same albumin chemistry and, thus, belonged to a single species. Needless to say all phases of this research—field work, acoustical experiments, morphological analysis of 1,500 specimens, blood assays—required extraordinary discipline and patience. Defining a problem in 1974, Heyer spent nearly seven years to solve it. Why? "I enjoy doing it," he says with simplicity belying a scientist's analytical sophistication. That's only part of the answer, which starts to come clearer in the Botany Department.

One needn't go so far afield as the amphibious realms of South America to grasp the possibly surprising fact that ignorance—compounded by some confusion—exceeds scientific understanding of the natural world. Botany curator Stanwyn G. Shetler recently concluded that in the north temperate zones of this hemisphere a single species of harebell mistakenly had 17 scientific names because the plant displays unusual diversity and locally adapted features. Needless to say, he devoted years of study to do something more than resolve a nomenclature problem, namely he sought by studying one array of plants to reveal aspects of plant distribution and variation which apply to other floral groups. (The nomenclatural chaos regarding nearctic harebells pales beside that in the palearctic region, by definition north temperate Eurasia. In that part of the world botanical study has suffered from a form of simple nationalism: Until recently, investigators tended to work within one country's borders though few plants recognize political boundaries. As a result belated comparison of the harebells, presumed to comprise some 700 names, revealed a host of synonyms and reduced the number of true species to forty.

Meanwhile, Museum curators have been able to explore other botanical frontiers. Curator Mason E. Hale Jr. recently found lichens growing in habitats which science previously considered barren, and which Hale has only begun to explain: namely *within* Antarctic rocks. Closer to the equator the bamboos, grasses which must be counted among the world's most useful and varied plants, include species that bloom only once a century. And when a species finally comes into flower all of its kind around the world bloom almost simultaneously, as curator Thomas R. Soderstrom recently demonstrated. (Discovery of the mechanisms that trigger this florescence holds awesome promise for bamboo agriculture, habitat protection and control of these plants which can be virtually ineradicable until they bloom—and then almost immediately die.) While laymen tend to take botanical fecundity for granted, Ghana-born botany curator Edward Ayensu and the late Paul D. Hurd Jr., a Museum entomologist, have revealed crucial faunal connections. For example, Africa's nocturnal bats —not insects—are now known to be the sole pollinators of many native plants.

Reviewing these disparate examples of botanical study at the Museum brings up a common thread which deserves emphasis. They all involve taxonomy. By and large the scientists who performed these investigations call themselves taxonomists, but something more than an obvious tautology is at stake. The complex discipline of taxonomy serves as the vehicle that enables science slowly to discover and illuminate many aspects of the natural world.

If there is a larger reason, it involves almost an article of faith shared by all biologists regardless of their special interests. Clyde Roper expressed it most eloquently as head of the Invertebrate Zoology Department: "What fills up the world? Plants and animals. Therefore it behooves us to find out all that we can about them." Beyond that no rational person will deny that given the jungle-like complexity of the world's biota, the best way to study it is systematically in both senses of the word: with some sort of a system which, as it happens, involves taxonomy.

A MARINE MUSEUM

Little by little it changes, yet aspects remain. Next time you see it all may seem transformed, yet it is the same place it was: protean and perpetual, ever changing and always constant. So be it with the sea and the Sea Life Hall, a place that reflects some of the most remarkable discoveries about life on earth in recent years. As every beachcomber knows, the sea changes by the day and hour. The Hall changes by the year and month and before the decade is out may undergo a metamorphosis as distinct as the Dinosaur Hall across the Rotunda. In the meantime it gains a three-quarter scale model of *Alvin*; then the unique animals this remarkable research submarine brought back from miles below the Pacific; then a living coral reef, the first habitat of its kind to thrive separate from the sea itself. Yet this dim and fascinating chamber, one of the most perennially popular, houses a legendary old attraction too, a model cetacean: the blue whale. First things first.

For decades the great whale hung in static splendor, its distant back curved, starting to dive within reach of amazed visitors. Huge and benign, somehow a good-humored beast, it was for years the largest of its kind, an uncanny copy in plastic, Fiberglas and paint of the largest animal ever to inhabit earth. Arcing 89 feet along the north wall, it represented a species that once ranged every ocean on earth with individuals that reached over 100 feet, longer than the *Brontosaurus* and a few times heavier to top 300,000 pounds. The Museum's whale has been a wonder and a symbol both of this place and of the world's unfathomed mysteries, just as the living animal it depicts became a symbol of the human yearning to stop the mindless genocide that threatens so many animals and plants.

The model for this model was an animal that washed up on the English coast and died. A "royal fish" (the Crown's property once it came ashore), it attracted both sightseers and scientists who took its measurements as it lay dead upon the alien sand. The British Museum used these morbid statistics to construct a same-sized copy for display, then offered their data to the Smithsonian. Since the dead whale was smaller than many of its living kind, the Museum took the liberty of enlarging its dimensions to make this model, then the biggest in the world's museums. In turn, the American Museum of Natural History, the

This scallop from the Dry Tortugas stands silent as a clam while marine scientists of many disciplines slowly illuminate the nature of the sea, life's first habitat and still its most fertile environment.

Overleaf: *A life-sized model of the blue whale, largest animal that ever lived, stretches nearly the length of the Life in the Sea Hall. Silhouettes of other living marine mammals appear on the wall behind it. Only the submarine* Alvin*—a three-fourths model of the famous research vessel hanging beside the whale's flank and above two men—is not true to real scale.*

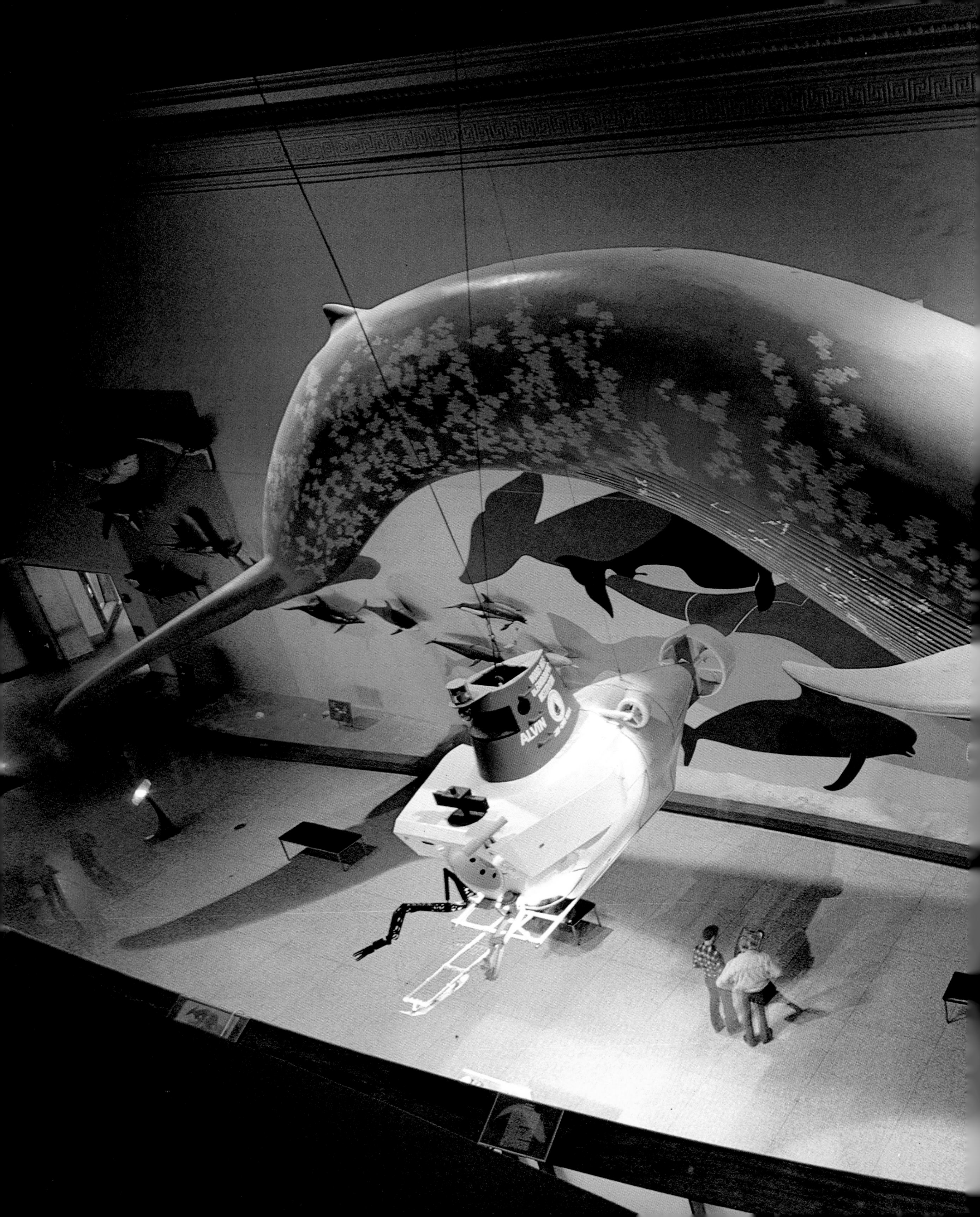
ALVIN

Opposite: *Marine animals show themselves off in a remarkable Museum exhibit (from top to bottom and left to right): a barracuda, butterfly fish, queen angel, brain coral, yellow tang, queen angel, black sea urchin, feather duster worm, short-spined sea urchin, squirrel fish.*

Pages 418–21: *The Coral Reef exhibit embodies the transplant of a living marine habitat and its inhabitants—a kind of mega-organism—from the tropical Caribbean seabottom to the center of a temperate and very terrestrial city. Developed from a smaller model in the Museum's basement, the exhibit represents a kind of Smithsonian symbiosis: While offering the public a mesmerizing display, it also serves as a self-perpetuating ecosystem for marine study, microcosmic observation and controlled experiments.*

The work leading to this reef represents the first removal of a whole marine community and its re-establishment apart from the sea itself. This 3,000-gallon tank contains those most fastidious invertebrates, corals, which have never been kept alive outside the sea before.

While new corals and other organisms are periodically added to the tank, the reef itself has grown as its resident corals act like they do in the wild. Thus far some 500 species of organisms have been seen in the tank, but that inventory may be incomplete since others undoubtedly lurk in rock and coral crevices.

Institution's friendly rival in New York, took the dimensions and enlarged them further for the model it constructed. Then all these mammoth mannikins were overtaken by scientific events. New interest, new diving gear and new divers' daring produced the first submarine photographs of whales in their natural habitat, the fluid ocean, which supports them differently from the way dry land does. (Axiomatically, the largest creatures ever to live did so in water, which supports and lightens more bulk than walking limbs could ever carry ashore.) With these new pictures, taken by photographers swimming among great whales, three well-respected museums experienced brief institutional embarrassment: their titanic exhibits now appeared wrong by comparison, each of them a deformed mock-up. But the embarrassment proved temporary.

Normally most blue whales appear slimmer and sleeker than the model when they swim, especially around the grooved throat which reaches nearly half the animal's overall length. They belong to the toothless suborder Mysteceti of the cetaceans (mammals that left the land to live in the rich oceans 50 million years ago); instead of teeth they have long fringed plates growing down inside their lips from their upper jaws. When these whales feed, they open their huge mouths to take in vast quantities of water and whatever small organisms swim in it—sardines, shrimplike krill, plankton. Then they close their mouths and force the water out through the sieve of baleen plates, catching their food in hundreds and sometimes millions of those small organisms. Part of the anatomical equipment for this extraordinary way of eating is an expansive throat that blows up like a pleated balloon. At rest the throat relaxes; the pleats collapse into "tubes"—or so early Scandinavian whalers saw these structures when they sliced into the animals they called "rorquals," a term derived from their word for tube. Well then, when a blue whale has half a mouthful, its throat remains partly distended; the animal only has the gracefully slim profile that the new photographers revealed between meals. Thus, some biologists suggest the Museum's whale isn't absolutely "wrong." It was just captured by the modelmakers with half a mouthful as it finished eating and began to dive.

From time to time, as plans progress for renovating the Hall, some

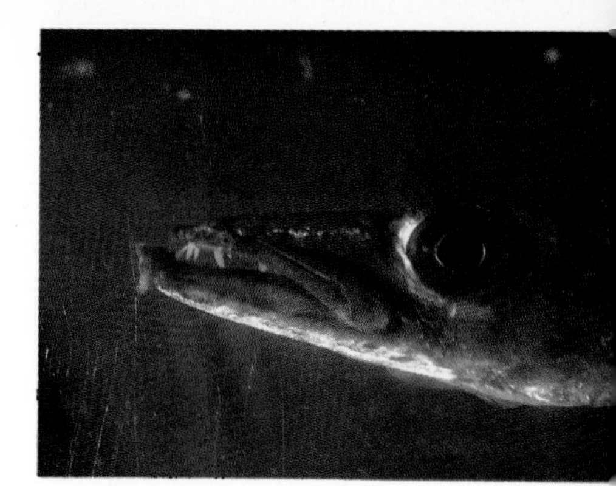

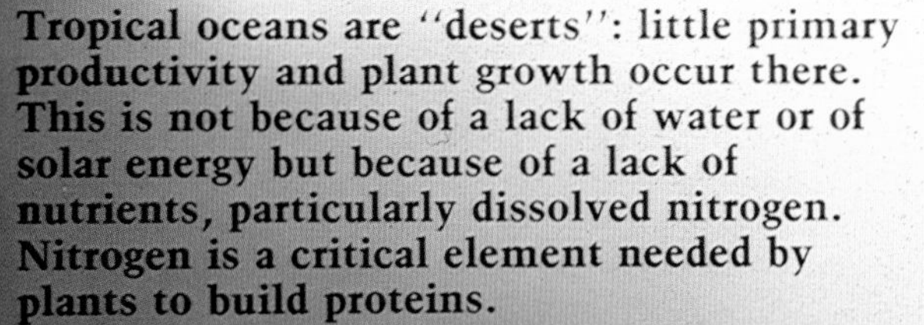

Tropical oceans are "deserts": little primary productivity and plant growth occur there. This is not because of a lack of water or of solar energy but because of a lack of nutrients, particularly dissolved nitrogen. Nitrogen is a critical element needed by plants to build proteins.

The coral reef exists like an oasis in the midst of this marine desert. The highest rates of primary productivity measured anywhere on earth occur in water from the open ocean flowing over the reef. To find out why, scientists travel to reefs to conduct experiments. As a result of these experiments, scientists have been able to re-create in the laboratory this living reef, which also serves as a tool for further research.

Open Ocean

Light level here equals light level at 60 feet [18 meters] on a natural reef

EXIT
CFA 200
KIMWIPES
scientific
0 20 40 60 80 100

HELIUM

Opposite: *From the other side of the marine world,* Murex cornucervi *(which translates as staghorn murex), an Australian gastropod.*

curators sadly suggest scrapping the whale. They reason that it's "not a specimen but a model—and not even the largest model any more." Worse, they note that it's not quite "right." But it's not all wrong either, and it does remain a spectacular object, an awesome thing deserving an early naturalist's reverent description of a sperm whale he witnessed fighting a giant squid as "awful"—inspiring simple awe in all of us.

So far as the inner Museum's work goes, the wonder on the wall represents the primary concerns of Dr. James G. Mead, curator of marine mammals, the latest taxonomist to continue work begun a century ago. In those times his research would have been easier, ironically, for he told me several years back that the best information about whale behavior and ecology lies in nineteenth-century records. The logs and journals of the whaling captains, and the narratives of authors who sailed with them, notably Herman Melville's *Moby Dick*, contain more numerous observations of these animals. There were, simply, more of them to observe. Thus Mead gathers new data where he can. In his youth he worked at a whaling station, about the only place he could encounter the creatures that fascinated him. In recent years he has studied strandings along the coasts of America, traveling hundreds of miles in a specially equipped flatbed truck and often bringing home the cetacean. (To size it up, he'd pull into a truck weighing station, a practice that startled some highway officials.)

Mead has found whale species along coasts where conventional lore said they never ranged, and he's defined migration patterns and schedules that Ahab would have given his other leg to study. But mysteries still outnumber the knowns. At one point he suspected that parasites in the ears of pothead whales disoriented the leaders of the great pods that came ashore periodically. But examination of more and more stranding victims convinced him that potheads have crustacea in their ears as commonly as dogs have fleas. To improve his data-gathering he helped establish the Scientific Event Alert Network, which collects information about sea mammal appearances, sea turtle sightings and geophysical events such as volcano eruptions.

George R. Zug takes special interest in SEAN's loggerhead turtle reports because of his long-term effort to learn their longevity. For

Opposite: *This sectioned* Fusinus *from the Indo-Pacific reveals a sea snail shell's construction. The animal starts life within the tiny apex, then grows in a spiral fashion, secreting new shell material and continuing to fill nearly the entire interior of its external skeleton.*

several years a collaborator has salvaged every turtle that washed up dead on Cumberland Island, Georgia, then sent the skull and right front flipper to the Museum. Examining microscopic sections of the flipper bone and comparing them with slides taken from turtles of known ages, Zug believes he can identify annuli—growth rings caused by the development of denser tissue during no-growth periods. Correlating ring patterns and the sizes of dead Cumberland Island turtles, he offers surprising news: loggerheads don't take 7 to 15 years to reach sexual maturity as expected, but probably 20 to 40 years. In part because of his findings, Federal authorities took steps to protect the species from intentional and accidental capture. Southern shrimp fishermen were reluctant to respect the new regulations at first, saying that the special panels that kept big creatures out of their trawl nets were an unnecessary expense. Then it turned out that by keeping turtles, big fishes and seals out of the nets, their catches sustained less damage from these hulks. The panels paid for themselves in a single voyage.

By happenstance, Dr. Zug is also the Museum's "monster expert," a respected scientist who has seriously studied the possibility that large vertebrates inhabit Loch Ness and Lake Champlain. Integrated sonar/photography studies suggest that something big lives in both lakes, although exactly what kind of creature it is remains to be clearly seen. An amphibian? Not likely, says Zug, because Loch Ness in particular began as a narrow inlet from the sea and no amphibians are known to have lived in salt water since their ancestors began foraging ashore. A worm? Nope, because the more reliable eyewitness reports and photographs suggest an animal about 25 feet long, much bigger than any of those organisms have ever grown—so far as science now knows. A mollusk? A fish perhaps? Possibly, or even a swimming reptile like the pleiosaurs, which were last seen 100 million years ago. For the time being "Nessie" remains an unknown, albeit one with a scientific name, *Nessiteras rhombopteryx* ("Ness wonder with diamond-shaped fin"), based on the computer-enhanced photographs. The intriguing thing in all of this lies in the consistency of descriptions by people who claim to have seen a monster both in modern times and antiquity. Ancient Scottish legends told of

Chosen for their beauty, a coral and the "harp of David" sea snail adorn the "Splendors of Nature" exhibit.

Following pages: *The size and shape of a chrysanthemum, this specimen of* Acropora spicifera *from the Caroline Islands belongs to the family of staghorn corals. The group got its common name, which hardly suits this species, from the antler-like appearance of its close relatives.*

water-bulls or water-horses in the highland lakes; the Indians whom Champlain himself encountered described similar beasts.

Once again, science must recognize greater ignorance than knowledge—a fact of life that should give us pause. It goes without saying that the best proofs of this premise come from the aquatic world, a part of earth we are only beginning to explore. Again, new technology and improved techniques are paving the way, and in the process revealing zoological phenomena of such native cunning as to make our complex tools seem clumsy.

Clyde Roper confirmed one such phenomenon in squid, which possess structures that selectively provide camouflaging light. Obviously any animal swimming above a potential predator during daylight hours would ordinarily be silhouetted against the downwelling light. But bioluminescence could hide the animal—if its "counterillumination" matched the brightness of the background. Dr. Roper went to sea in a research vessel and netted a number of organisms at night from 300- to 1000-foot depths. He placed them in a special tank in a pitch-dark shipboard laboratory. Varying the filtered lights above the tank to mimic intensities at different ocean depths during the day, he found that the squids adjusted the "photophores" in their ventral surfaces to match the apparent surroundings and become invisible from below. Later experiments showed the squids used more than their eyes to monitor what was going on. Lightproof shields placed over other light-sensitive organs on the ventral and dorsal sides of the animal's head demonstrated a feedback mechanism: the upper organ

monitored the light level from above, while the lower organ measured the light emanating from the squid's own photophores.

Perhaps even more surprising, an anglerfish displayed similar ability. When this animal was brought aboard and placed in a holding tank on deck, Roper was standing at its head and his collaborator at its tail. The collaborator said, "It's bioluminescing," and Roper thought him crazy until he walked around the tank. The fish was shining astern—assuming that anglerfish swim face forward like "proper" fish. But that isn't the case; when they put the fish in the darkroom tank, it oriented itself vertically with head up, and all its radiance shining downwards. (Anglerfish have some unusual traits. Many species habituate great depths where conditions are so relentlessly dark that individuals presumably encounter potential mates only by accident. When they do, the male attaches himself to the female by his ample jaws and holds on for dear life and posterity. His tissues fuse with hers; even circulatory systems unite since he has lost the ability to eat for himself. Only his reproductive organs remain functionally independent so that when she spawns her eggs he follows suit with the sperm-laden milt. Roper's anglerfish, a female, had a little parasitic male attached and he glowed in the right direction along with his hostess.) Adjusting the tank's light source, Roper watched the fish(es) match the downwelling light's intensity. When he raised the light to mimic intensities in shallower water than this species occupies by choice, it frantically tried to swim through the bottom of the tank and became comatose in a fit of photophobia.

There seems to be no limit to the kinds of relationships that marine organisms enter. A number of seaweeds spend a major part of their metabolic energy creating compounds toxic to algae-eating animals. Presumably this represents a straightforward defense mechanism against predators. But the animals, in turn, have developed counter offenses in what Jim Norris calls a floral-faunal arms race. Fishes that "graze" submarine plants off Belize habitually avoid the poisonous seaweeds, while "browsers" nibble a little from one species, and a little from another, apparently avoiding too much toxin of any one kind. Then there are the species-specific nudibranchs (mollusks of the sea hare and sea slug family).

Opposite: *Marine biologists cannot work exclusively in musty landlocked laboratories. Carrie Bow Cay, off the coast of Belize, houses a small research station as headquarters for Museum scientists' long-term studies of Gulf of Mexico flora and fauna.*

Pages 432–33: *An amateur collector arranged this constellation of brittle stars.*

Pages 434–35: *Off America's southeast coast this porcelain crab practices peculiar, not to say pushy, living habits.* Porcellana sayana *will move right into a mollusk shell already occupied by a hermit crab. Presumably the lodger benefits from the defenses his host provides. What the hermit crab gets out of this odd cohabitation remains to be seen.*

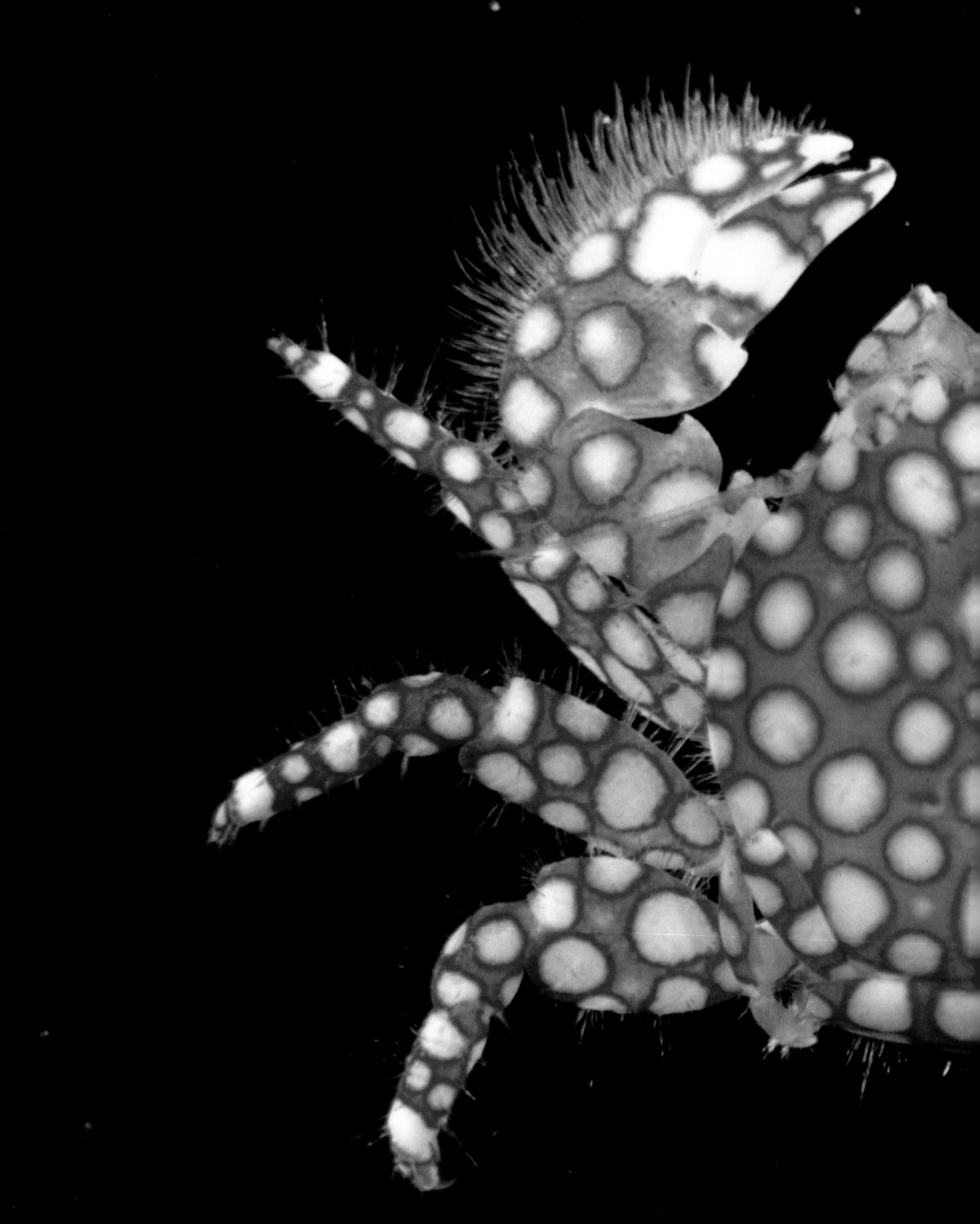

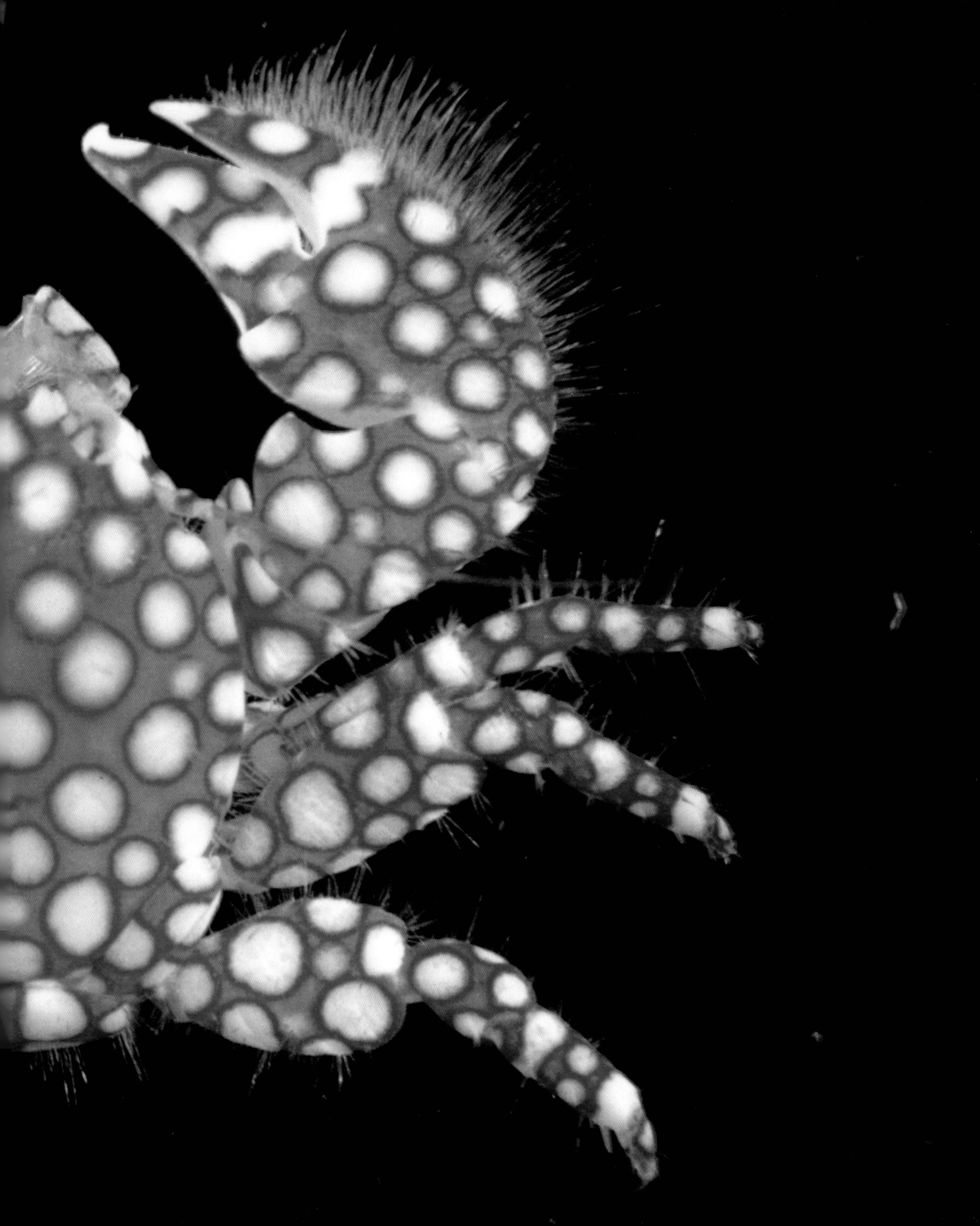

Opposite: *Noted for a lovely exterior,* Conus textile *possesses remarkable internal equipment to support a predatory life-style. The little sea snail extends a "telescoping" appendage—actually its feeding tube—within reach of small fishes, worms or other mollusks. It stabs its prey with a poisonous dart which leaves its own body, and eats the fallen victim at leisure. This Pacific denizen has toxin strong enough to have caused human deaths.*

Norris found that one *Aplysia* species feeds almost exclusively on *Laurencia*, a particularly poisonous red alga, and survives by extracting the toxin in special digestive glands. The nudibranch not only thrives on the weed, it eats nothing else. Further, nothing else eats it; its freedom from predation has been a widely observed phenomenon for years. Presumably the reason lies in its chronic intoxication with the alga's poisons. What limits the nudibranch's population then? A kind of invertebrate self-control. Other experimenters studying the animal were stymied in their work because, while they could raise *Aplysia* larvae, they could never induce them to go through metamorphosis in captivity—until someone thought to spice the water with *Laurencia* extract. Traces of the adult food source triggered the mature stage. (These examples of co-evolution have interesting tangents. To test the toxicity of various algal defense compounds, Dr. Norris exposed common bacteria to them—staphylococcus germs, the organisms that cause serious infections in humans and might respond to drugs formulated from the seaweeds. More dramatic, one of the algal toxins stops division of sea urchin eggs in their tracks. If the compound can inhibit the growth of these most prolific cells, might it be used to retard cancerous growths?)

Tangentially, it bears mention that the lowly nudibranch's adaptive cunning belies the commonplace notion that some animals, namely us, are intrinsically superior to them. The reason for this—or its symptom—lies in the relative terms "higher" and "lower." But zoologists wisely use these words to describe various organisms without the odious implications that many lay people infer. The reasons for the scientists' lack of prejudice seem plain enough; they find "primitive" fauna (to use another misleading adjective) at least as interesting and complex as supposedly cerebral man. From many perspectives the achievements of humans pale beside those of other beings. Given almost any single standard, other life forms excel biologically beyond the wildest dreams of *Homo sapiens* and this holds true whether one considers aquatic, terrestrial or ethereal animals. Corals build more extensively, whales have bigger brains, spiders exhibit finer agility. Among the birds, hawks see several times more acutely than we do and terns cross oceans without compasses or fossil fuels. Turtles live for centuries. Moths see into the ultraviolet. Cockroaches avoid eating poison

by taste-testing everything with their feet. Without bothering to change in any significant way, horseshoe crabs and rainbow trout have gone about their productive business far longer than any mammal. Many insects adapt faster to new conditions than any vertebrate, let alone faster than the "highest" and most prodigal predator can invent synthetic repellents. The list goes on. People have unique talents, of course, abstractly creative abilities for the most part. We communicate between generations, sending intangible lore to the unborn and reading messages from the dead; we invent and replicate symphonies, bronze sculptures and chocolate mousses unlike anything that ever appeared biologically. Curiously, we deduce information about the nature of nature.

The fundamental biological fact remains that every organism "down" to the most rudimentary virus performs every significant organic function that man performs. Often individual animals prevail against odds of many times greater magnitude than any human has encountered since the dawn of agriculture. In the macro- or microscopic wild, animals without hands must feed themselves—find food, ingest it, absorb selected constituents and expel others with their own metabolic wastes. Nearly every nonhuman creature competes for life with organisms many times its own size and strength. (If one happens to be among the largest in a habitat, like the bowhead whale or water buffalo, it must contend with pods of orcas or prides of lions that collectively out-weigh and out-bite it.) Animals without feet, feathers or fins routinely travel enormous relative distances to find safety, breeding partners and brooding sites. All living creatures experience the trauma of birth, or hatching, or budding like the hydra, without the benefit of midwifery. They routinely survive countless deadly hostilities—at least some individuals of every species must, or their kind has no future. All transmit a multitude of traits and characteristics to their progeny. In these respects and others, we have no intrinsic superiority; the sand flea and the sea hare are every bit as productive as any primate with a high school diploma. So what might be meant by the invidious relativity of "higher" versus "lower" animals in the lingua franca of zoology?

It involves something beyond simple semantic convenience, though

Opposite: *The Comb of Venus* Murex, *a gastropod of unusual beauty chosen for display in the "Splendors of Nature" exhibit, bristles with an unanswered question: What purpose do its spines serve? One malocologist suggests they keep the animal from sinking in the thin mud of its Okinawa habitat. Others believe they offer protection from predators which naturally might prefer morsels that aren't so hard to swallow. The difference of opinion reflects a simple fact: Science has barely begun to understand the sea and its denizens.*

Overleaf: Spondylus americanus, *the spiny oyster of tropical America, commonly goes by the name chrysanthemum shell. Appearances aside, it is only distantly related to the twin specimens of* Tridacna squamosa, *which grows to a foot or more. These are the dwarfs of their genus, which includes* T. gigas, *an Indo-Pacific reef clam that grows to four-and-a-half feet. Curator Joseph Rosewater, a world authority on these bivalves, reports they inhabit shallow water and open their shells in dappled sunlight, allowing internal algae to photosynthesize and grow. Then tridacnas "harvest" the algae in a relationship resembling that of the newly discovered riftian worms and their internal bacteria.*

Opposite: *The "Glory of the Sea Cone's" name translates exactly from the Latin* Conus gloriamaris. *This one was taken in Magellan Bay, the Philippines.*

verbal economy might have been the first cause. The useful distinction relates to phylogenetics, not that the fruit fly or bottle-nosed dolphin has evolved any less than our closer cousin the chimpanzee. (Indeed it might be argued that over time monkeys sustained less dramatic changes than cetaceans. Ancestors of the latter left the aquatic world, adapted to the terrestrial one, then went through the arduous business of going back to sea and becoming whales.) Relative elevation in terms of taxonomic adjectives usually does reflect a particular distinction in this respect: a "higher" animal, a vertebrate for instance, belongs to a phylum that appeared later or branched off farther up the trunk of the tree of life than that of a "lower" earthworm. All living material—in animals, vegetables and forms with a root/foot in either kingdom—is composed of different sequences of the same 22 amino acids, the common components of all mundane life. This fact demonstrates that all forms derive from a single beginning. (If other vital sources appeared, they evidently lost the competition for living space.) Thus each one's ancestry might be hypothetically traced back to the same single moment of protoplasmic genesis in the primordial past. But the intermediate ancestors of some "lower" organisms like the sea cucumbers diverged from the common stock into distinct general forms before every "higher" class of critter with a central nervous system. They developed a formal body plan and a particular way of living that has served them well for upwards of 700 million years. Long before the age of Bumper-Sticker Wisdom they intrinsically understood the watchword "If it works, don't fix it."

Zoologists might have settled on the relative terms "earlier" and "later" forms of life to make a commonly useful distinction if it weren't for the fact that bacteria, for example, can spin off new species faster than rabbits multiply. Thus some minute organisms—the one responsible for the latest brand of flu—are younger than anything with hair. Zoologists might have hit on the relative terms "primitive" and "advanced," though what can be more remarkably sophisticated than a sea hare's using its preferred food's poison to make itself unpalatable to potential predators?

Be all that as it may. Other pairs of adjectives might have been used but each set has its own shortcomings. Meanwhile, "higher" and "lower"

Opposite: *Like a locomotive palm tree, a sabellid worm, removed from its tubular home in the sea floor off Florida, wiggles away in search of safety.*

Following pages: *Its magnified "radiolis," normally used to ensnare plankton food from the water, resembles ferns or feathers.*

are the modifiers that stuck in the scientific lexicon. No offense to the myriad spineless ones is intended, and certainly none taken as they pursue their endlessly complicated and fascinating yet pre-cerebral lives. No arthropod or annelid ever developed a complex brain, which would only have been excess baggage. They do superbly well without them.

One of the most striking and humbling examples of the adaptive superiority of "primitives" lies deep beneath the Pacific. There, beyond man's reach or imagination until quite recently, complex communities of organisms have been thriving for inestimable time absolutely free of the photosynthetic world. This peculiar independence in itself seemed beyond the realm of biological probability—until *Alvin* took scientific curiosity to new marine depths. This research submarine, 15 years old at this writing, arose from a kind of bureaucratic interaction almost as complex as the benthic regions it explores. Owned by the Navy, and operated by the Woods Hole Oceanographic Institution, the sub is supported largely by the National Oceanographic and Atmospheric Administration and the National Science Foundation. Named for designer Allyn Vine, the machine has proved a versatile seahorse capable of carrying a pilot and two passengers to depths of two-and-a-half miles. Within two years of launching, the Navy used her to recover a hydrogen bomb lost in the Mediterranean off the coast of Spain. Later she herself would be lost at a depth of 5,000 feet for ten months after a cable snapped and her crew scrambled to safety. Despite being trapped for an hour in an Atlantic bottom crevice, and nearly being hit by a benthic landslide near the Galapagos, she has made more than 1,100 dives—most of them on research missions—and continues to chalk up 100 a year. Her two priceless talents: taking landlubberly scientists on observation cruises, and manipulating things on the bottom, then bringing back samples of the substrate and its biota.

It was among the Pacific rifts—seafloor spreading zones where tectonic plates slowly separate—that she crossed a new biological frontier at a depth of a mile and a half. Every schoolchild knows (or knew) that the sun provides all the energy for life on earth, that its light enables plants to make organic food out of inorganic minerals through photosynthesis. Most

Opposite: Penaeus marginatus, *a congeneric relative of the culinarily familiar brown, white and pink shrimp caught off America's southeast coast, ranges the distant Pacific from Hawaii to Africa. It is one of 300 species prized by the world's commercial fisheries, but 2,200 additional species inhabit the world's oceans.*

plants are autotrophes (from the Greek for "self" and "nutrition") in that they make their own organic fuel. Animals (heterotrophes) rely on sources outside themselves for food. Some carnivores live on other carnivores, as falcons eat songbirds that eat insects, but every food chain leads down to vegetable material. (The insects that birds eat live on leaves and such or on other insects that do.) Every food chain, that is, except those of the rift zones—which now appear to depend on at least 40 strains of bacteria that seem to derive their energy from hot springs in the rifts themselves.

At this writing a complete description of the rift zones' ecosystem remains to be compiled. But this much can be said: each of these regions explored so far is a submarine island of sorts that receives only negligible amounts of organic material from the surrounding sea. Each is also a thermal maelstrom of sorts, buffeted by cold water hovering a couple of degrees above the freezing point, then swiftly heated by upwelling hot water from fissures in the rock. (One vent was hot enough to damage *Alvin*'s temperature probe.) More typically, waters of temperatures varying by as much as 70° C swirl and eddy among the rocks, subjecting organisms not only to hot and cold, but to high levels of sulphur compounds instead of oxygen.

This environment does not seem otherworldly to some scientists who believe it might resemble conditions that were typical when the world was young and life just beginning. Whether a throwback to the primordial seas or a persistent remnant, the perplexing regime supports communities of animals unlike anything seen before, some of them possibly primordial survivors, some the specially adapted descendents of highly advanced organisms. *Alvin* has returned with giant white-shelled clams and mussels the size of soup bowls. Austin B. Williams, a decapod taxonomist attached to the National Marine and Fisheries Service taxonomy lab at the Museum, has described a blind crab that experiences more larval phases than any of its kin. Presumably this animal descends from a sighted decapod, because it has eye stalks which, as in other crabs, contain vital hormone glands. But the eyes atop them do not develop in this denizen of a lightless environment. Thus, a seeing crab colonized to the riftian depths and abandoned its useless eyes but retained the still-essential parts that

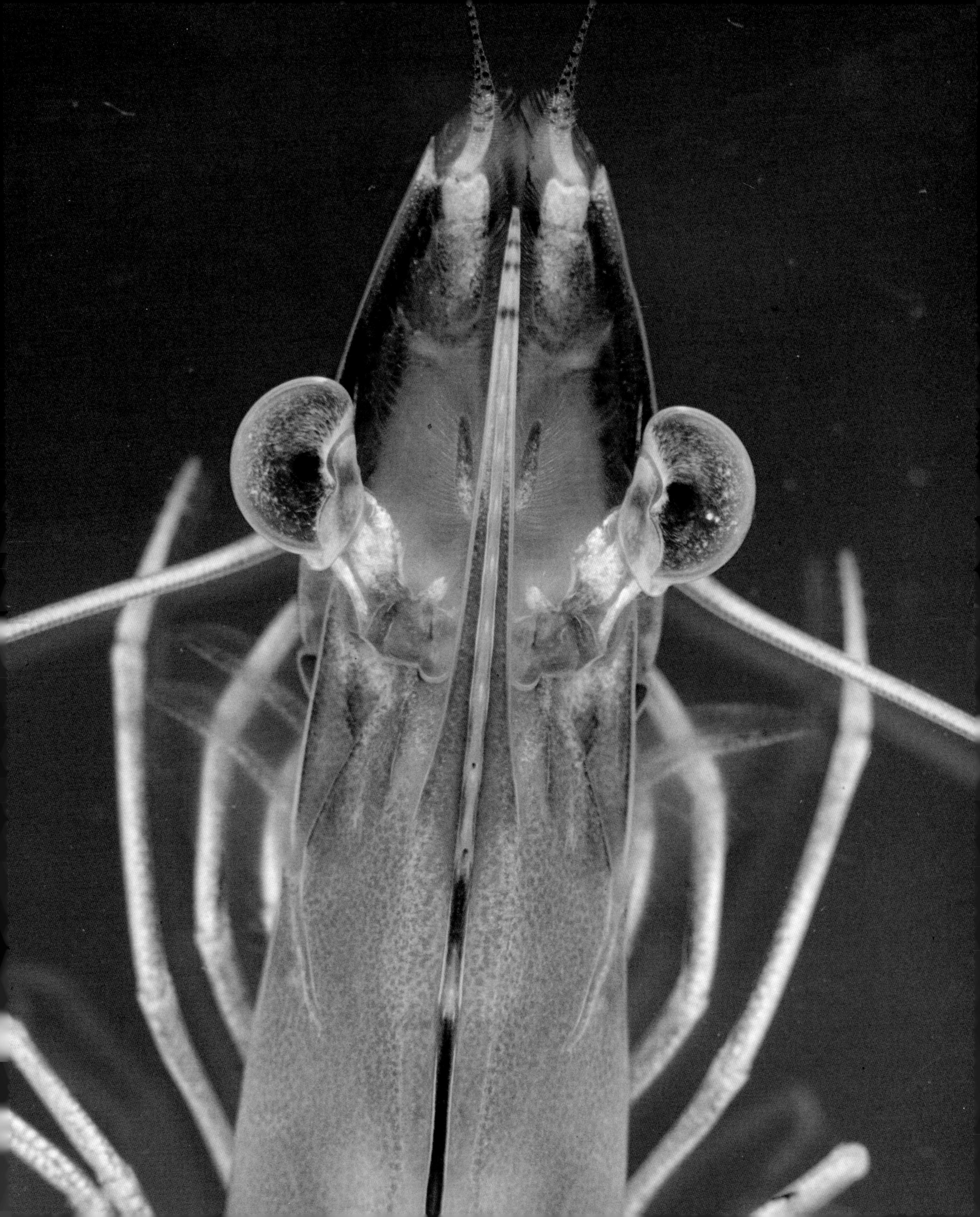

supported them. More unusual than the mollusks and crustacea are the giant worms, creatures that grow to five feet long in dense colonies in a marine soup thick with bacteria. Miraculously, it seems at first blush, these crimson-plumed worms have neither mouth nor gut. Meredith L. Jones, a worm taxonomist in the Invertebrate Zoology Department, discovered this unique anatomy and has begun to decipher the metabolic mechanisms that explain it.

First Dr. Jones addressed himself to the task of identifying the creatures (which he observed in their habitat during one of *Alvin*'s many dives). He decided they belong to a phylum whose first representatives were encountered only in 1914 and were not assigned to a distinct group until 1944. Their muddied taxonomic history reflects the complexity of vermiform biology. Fully a dozen of the 26 animal phyla recognized today represent long spineless things: flatworms, round worms, spiny-headed worms, spiny-skinned worms, scaled worms, priapus worms, peanut worms, sausage worms, horse worms, arrowworms, acorn worms and more, each group descending from a different primitive ancestor that left no trace in the beds of Precambrian seas. When a weird benthic worm was found off Indonesia in 1914, it was described as an "aberrant polycete," one of the annelids that have included the earthworms and bloodworms since Jean Baptiste de Lamarck named the group during the Napoleonic era. The oddity was left to languish in a jar of formalin. Twenty years later some similar specimens—inches in length and on the order of 100 times longer than wide—were dredged up from the muck off the Kamchatka Peninsula. Their captor hedged the taxonomic bets further by calling them simply "aberrant annelids." Another investigator backed off so far as to call them "aberrant worms," a term so vague as to be taxonomically meaningless. But they began to be of wider interest when it became apparent that similar organisms inhabit most marine sediments deeper than 600 feet.

By 1944 a Russian taxonomist found that the relative positions of the nerve cord and apparent gut were the reverse of those found in annelids. He declared them representatives of a new phylum and the latest to join the hierarchy, Pogonophora. Jones found their giant relatives from the Pacific rifts different in such significant ways that he assigned them to a new subphylum, then named the gutless wonders *Riftia* (for their habitat) *pachyptila* (for its "thick feathered" plume, a structure of remarkable dimensions). The plume divides into about 225,000 tentacles less than an inch long, each of them narrowing down to a single-cell thickness. Given the total surface area of these flat lamellae, the plume appears analogous to the gas-exchanging surfaces of vertebrate lungs or gills. But the question as to their respiration function was eclipsed by a more pressing one: if the animal lacks both mouth and gut, how does it eat? Thus far Jones and a team of far-flung collaborators have only been able to pinpoint the internal source of the nutrition: masses of sulphur-using bacteria contained in the body cavity. Dr. Jones noticed white spots in his specimens and had the particles analyzed by instruments in the Museum's Mineral Sciences Department. The electron microprobe showed them to be pure sulphur. Mentioning the results of this assay at a seminar on the worms he gave at Harvard, the taxonomist was intrigued by a doctoral candidate's intuitive guess, which proved correct: that the material surrounding the sulphur is bacterial. Further experiments showed the pure sulphur spots are byproducts of a metabolic process in which the bacteria use heat energy from the vent waters to convert inorganic carbon from the sea into organic carbon bound up in carbohydrates—the primary step in biochemistry that supports this deepsea community.

So here is an ecosystem that doesn't need photosynthesis, a remarkable phenomenon containing a host of still inexplicable mechanisms: how did the bacteria reach the rift zones that are widely separated by hundreds of miles in frigid water? How do the bacteria get inside the worms, which have no orifices besides those used for reproduction? Finally, what is life doing in those depths beyond the reach of sunlight, presumably the vital sine qua non? Could it be that the first life on earth was energized by the planet's own inner heat and not by the sun? One doesn't hear that hypothesis often in scientific circles, but then as awareness of our ignorance grows, we may be just beginning to ask better questions.

A MUSEUM OF MAN

The Olmec culture, "mother civilization of Mexico," produced this votive axe circa 1200 B.C. Less than nine inches tall and mounted on its nominal cutting edge, it bears an oddly fetal visage below a split, tapered brow. Its purpose—no doubt somehow ceremonial—and its symbolism have not been explained.

An archaeologist studying Eskimo and Indian ruins in subarctic Canada uses a biological metaphor to explain his fascination. Different cultures, he says, seem like distinct species in a genus. This is not to propose new taxonomic divisions within *Homo sapiens*; it is to assert that as new knowledge of one butterfly deepens understanding of its biological family, so discoveries about one clan increase knowledge of the human race. Just as the monarch butterfly's migration relates to the viceroy's behavior, so the taboos, rituals and ingenuity of one culture can illuminate another's ethos. This is to declare in scientific diction John Donne's poetic intuition that "No man is an Island, entire of itself." Elsewhere in the Museum a physical anthropologist looks up from a 10,000-year-old skull to suggest that modern man differs little in important ways from our Neandertal ancestors. Elsewhere still, an archivist studying the scarce records of Indian tribes writes, "It is difficult to understand the present without a thorough knowledge of the past, warts and all." Among these views—the first perceiving vast cultural diversity, the second a certain biological constancy and the third the value of historical mirrors—lie kaleidoscopic images reflecting distinct aspects of man's nature, the single enormous concern of the Museum of Man.

As it happens, the Museum building bears the bronze legend "National Museum of Man" along with the older name "National Museum of Natural History." (The newer title represents a hope of combining various independent programs within the Smithsonian under a single organizational umbrella and, in the distant future perhaps, under one new roof. But this union remains an unfulfilled ambition at best, indeed a bureaucratic fiction of sorts.) Humankind remains a primary concern here, a focal point identified by the first generation of Smithsonian scientists and perennially examined ever since. Yet the question returns: why continue to include anthropology within a natural history museum? The simple answer: because mankind is, first and foremost, a biological species that has lived nearly all its collective life in purely natural circumstances. We arose from the natural world and our roots remain there. We are part and parcel of natural history.

(It goes without saying that lately our role on the earthly stage seems

Opposite: *The hides of 14 buffalo cows were laced together into this Arapaho teepee, displayed at the Philadelphia Centennial Exposition of 1876. It may be the only surviving shelter of its kind.*

to have been particularly villainous in several ecological respects. A longer view reveals that hominids have always played such parts. During the Pleistocene Epoch, which ended 10,000 years ago, our increasingly predatory species may have irrevocably changed the fauna of all Eurasia. Subsequently on this continent, it is thought that proto-Indians pushed a number of big mammals over the brink of extinction. The waxing and waning of glaciers during a period of climatic vacillation had been hard enough on temperate-zone animals; emerging man just iced the proverbial cake. Whenever a group of species disappeared, a new balance emerged among the inheritors, just as a new balance—dominated by the growing variety of ever-larger mammals—had come into being when the dinosaurs died out.)

In other institutions, of course, new branches of ultra-modern anthropology examine purely man-made phenomena, such as the psycho-social dynamics of money. But the human sciences still include more firmly established courses and these are the ones that Museum scientists pursue: Archaeology, the examination of extinct societies through their buried artifacts; Physical Anthropology, the science of human evolution, the study of human populations, the use of physiology per se to illuminate ancient cultures and modern mysteries; Ethnology, especially the study of nominally "primitive" peoples who continue to live closer to earth and sea than industrial nations. These overlapping disciplines in turn comprise several other specialties such as linguistics, paleobiology, osteopathology and many more. All these sciences focus on specific threads within the huge tapestry of human experience; all the Museum scientists contribute to a common effort. In the down-to-earth words of one learned curator, "We're trying to make some sense out of people."

As a Museum department, Anthropology has more professional members than any other. Furthermore, fully a third of the Museum's public halls display material that is *ipso facto* anthropological, which means this department's range of behind-the-scenes work gets exhibited more comprehensively than the others'. Here stands the largest hide teepee remaining from the days when buffalo ranged the plains and *Bison bison* provided people with almost everything they needed in material terms. Backstage,

Opposite: *Imprisoned at Fort Marion, Florida, a Cheyenne warrior named Making Medicine drew "On the War Path" with crayons in 1875. The horsemen wear appropriate regalia: fringed warshirts and feather bonnets denoting previous bravery in battle, lizard amulets and sacred shields. The style reflects an ancient tradition among Plains Indians, who decorated teepees, drums and other animal-hide articles with epic pictures. "Ledger art," as such drawings came to be known for the paper often used, provides important ethnohistorical information. Making Medicine, later baptized David Pendleton, became an Episcopal minister and served among his people until his death in 1931.*

Highlights of the Battle of Greasy Grass River—from the victor's viewpoint—illuminate this page and the next two spreads. Five years after the fact a participant, Chief Red Horse, filled 41 sheets of ledger paper to illustrate the account he narrated in sign language. The first picture shows the defenders' encampment before the battle. The second depicts a shambles of slain horses—all of them enemy mounts (since Indians tied up their horses' tails for battle). Pages 458–59: *Red Horse's comrades repulse the invaders. Next* (pages 460–61), *a single picture of two apparent parts shows the hoofprints of U.S. Cavalry on their way into battle, then a closing phase of the rout as four soldiers flee back the way they came. In the end all cavalrymen were slain by Indians under Chief Sitting Bull's command in what is also called "Custer's Last Stand" or the Battle of the Little Bighorn, the whites' name for the Greasy Grass. As the Confederacy and the Union knew one place as Manassas and Bull Run, respectively, during the Civil War, adversaries in the American West had different names for single engagements and viewed landmark events diametrically. Documents like these in the Museum's National Anthropological Archives preserve many distinct views and provide leaven for modern assumptions about "objective" history.*

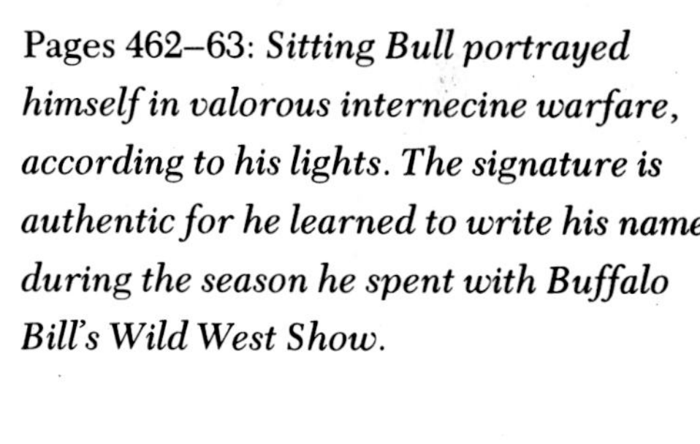

Pages 462–63: *Sitting Bull portrayed himself in valorous internecine warfare, according to his lights. The signature is authentic for he learned to write his name during the season he spent with Buffalo Bill's Wild West Show.*

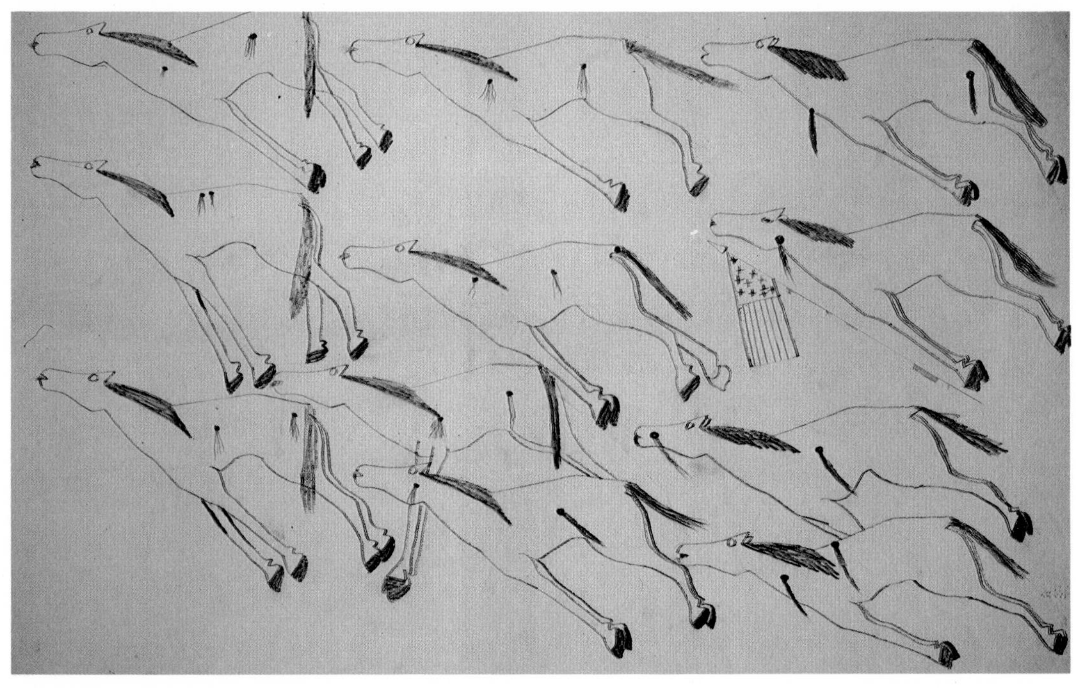

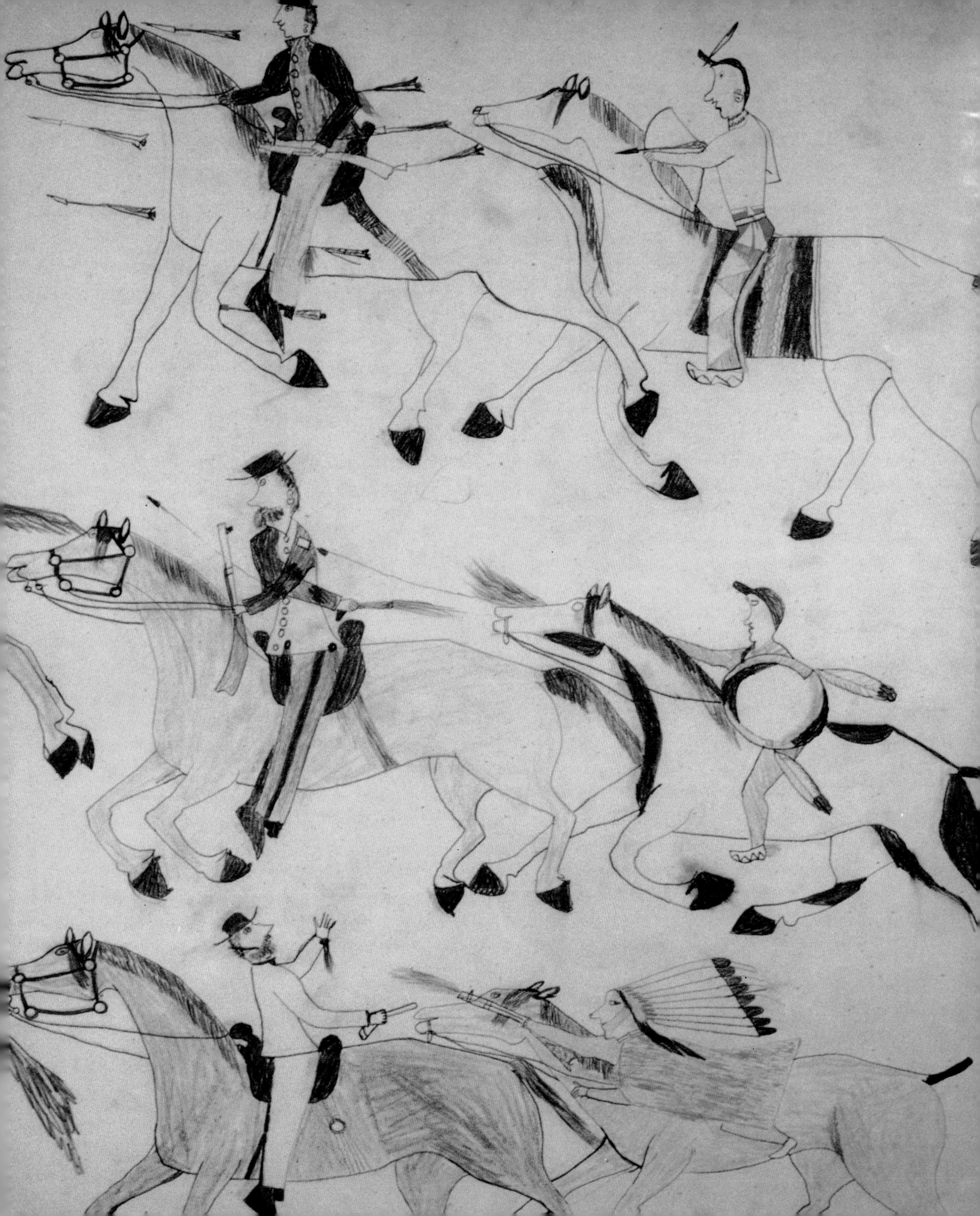

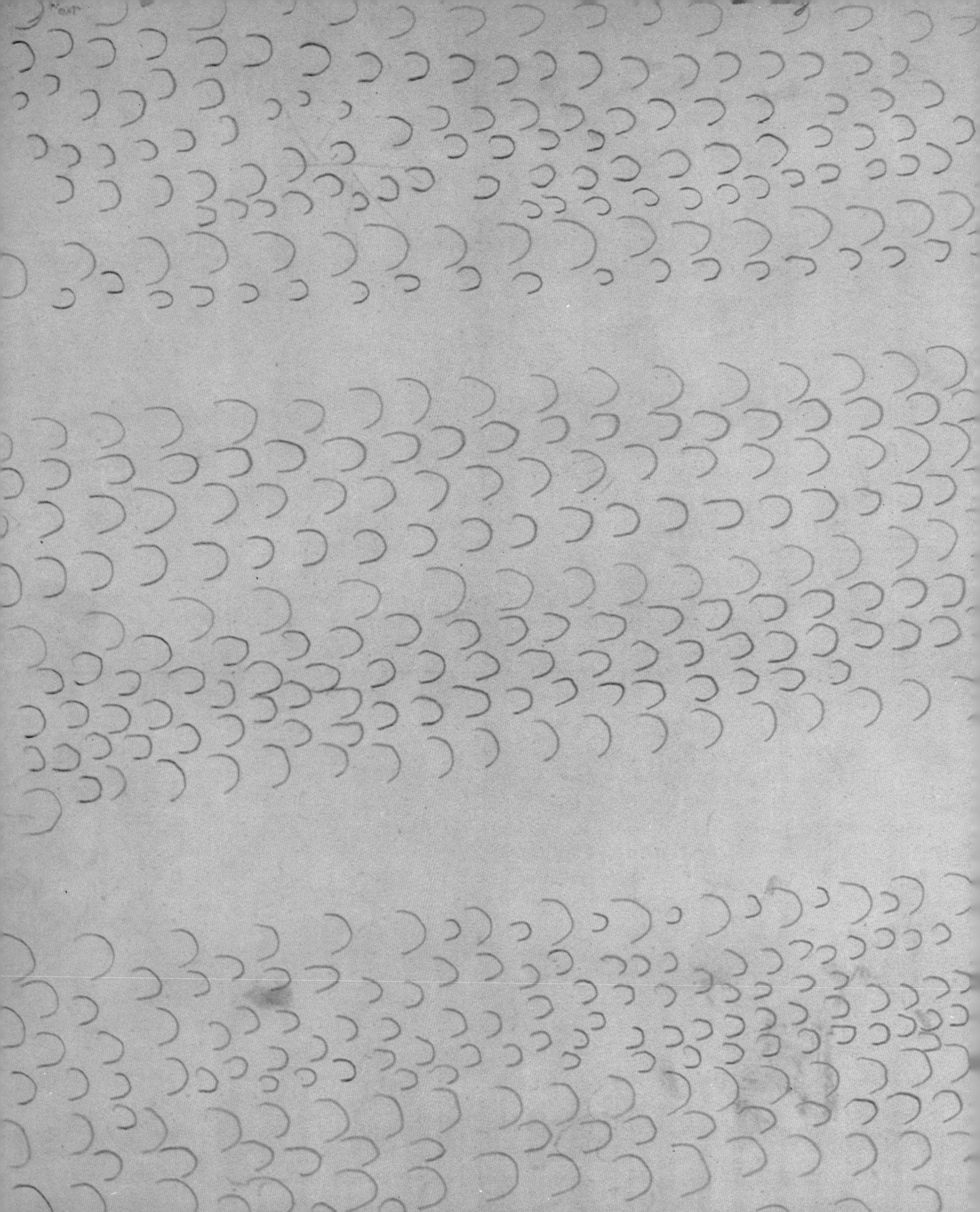

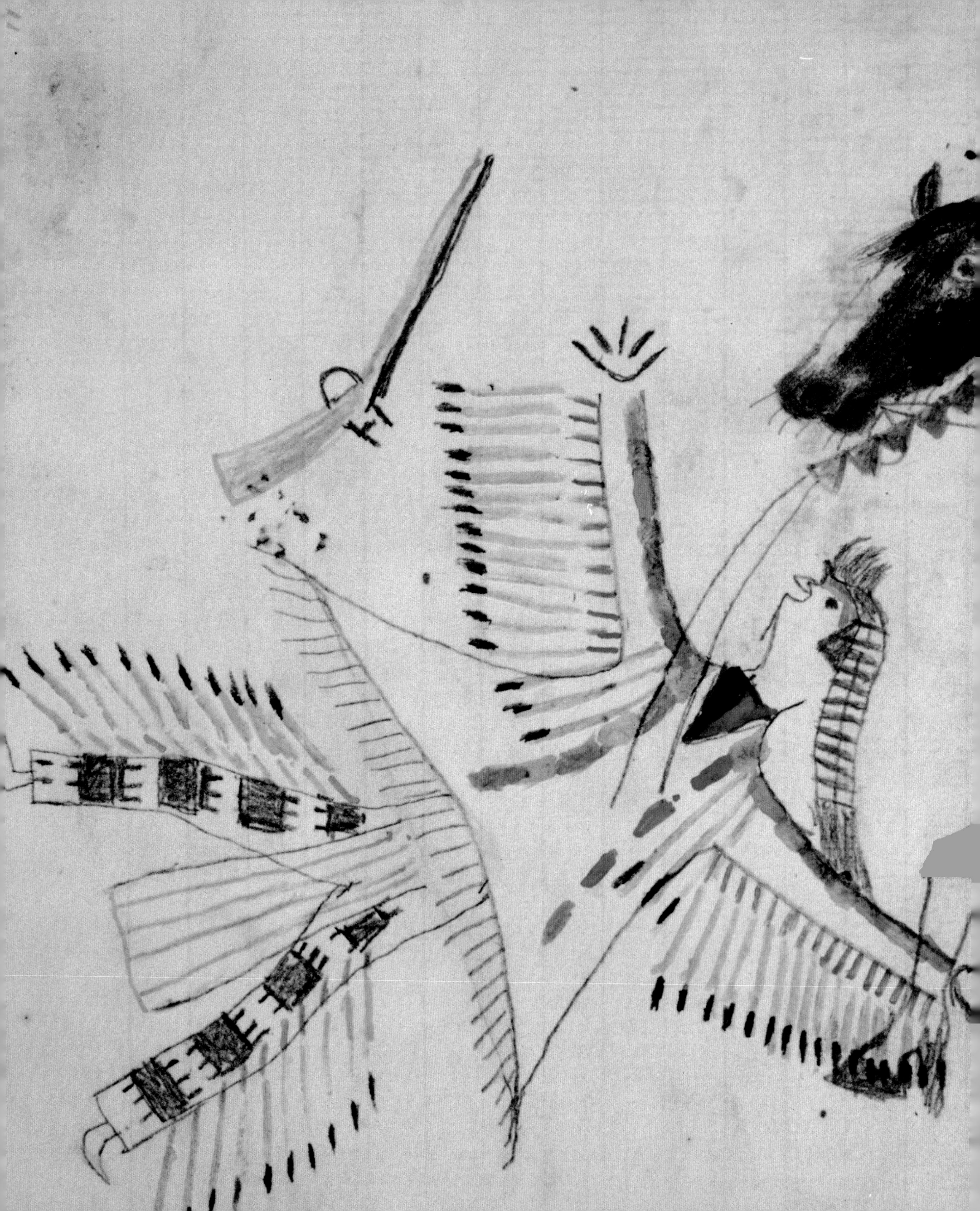

Sitting Bull

Opposite: *The famous Dan figure, possibly a fertility symbol, wears adornments and facial scars admired by the West Africans who made her out of brass more than a century ago. Cast by the lost wax method, she began as a model sculpted in beeswax, which was covered with clay. The clay mold was dried and heated to melt the wax, which flowed away. Molten brass was then poured in to fill the space and solidify as a replica of the wax form. The mold was broken, revealing a one-of-a-kind figure since the original wax and the clay form were gone. Importing copper from North Africa, West Africans became skilled in working it before trade with Europeans began in the fifteenth century.*

Pages 466–67: *With locally forged iron, Central Africans raised weapon design to a utilitarian art. The shape of each knife declared its tribal origin. From left: Yans, Ngala, Saka, Bai, Ntomba, Kuba, Kusu, Nktutshu. While the spear was meant for both hunting and fishing, smaller knives were designed for common tasks. The larger knives, used for sacrificial beheadings, were also carried as symbols of authority.*

Pages 468–69: *The !Kung (Bushmen), ethnically distinct from other Africans, range the Kalahari Desert in small communal bands. The woman at left makes beads from a broken ostrich eggshell, which has already provided food and may have served as a water canteen. The dancing child wears cocoons as leg rattles. From a seed-shell pot, the squatting man daubs an arrow shaft with lethal poison made from the larvae of a single beetle species. Until recent decades, at least, the Bushmen lived entirely by hunting and gathering wild foods.*

ethnologists and archivists alike study Indian cultures in depth. There is the sculpted figure of a fertility goddess, a priceless relic of the Dan people of West Africa, on display for the public to ponder; in distant corners of the edifice specialists study many African groups in detail. There are masks from everywhere, a commonplace among many cultures, perhaps a universal way of mimicking the features of perceived gods and honoring spirits captured in native materials such as wood, bone, feathers and clay. Indeed, the exhibit halls and the hidden collections they represent contain cultural artifacts and physical relics of every distinct human group known to the modern sciences.

If the Anthropology Department stands first among equals, it is also clearly the one which depends most on the expertise of other disciplines. A physical anthropologist studying skeletons in an ancient cemetery needs precise biochemical analysis to discover the bones' special characteristics. An ethnologist examining the wicker baskets of extinct Pacific islanders relies on botanists to identify the pollen grains preserved within their weave in order to list the plants the people knew. Linguists now use computers to quantify differences in verbal elements and thus determine how far two languages have developed from a single source. (Here, the branching of tongues resembles biological radiation.) Archaeologists employ geological stratigraphy and radiocarbon dating to gauge the age of different soil strata as they proceed down through the ancient layers of a dig. They need metallurgists to assay bronze tools and pinpoint the source of the ores from which they were made. They consult zoologists to determine which animals a society raised for food—as revealed by midden debris—and which wild beasts may have shared the territory. In sum, anthropology seems the most catholic of the sciences, which may explain why the Museum's curators have collected everything from Jordan's Bronze Age funeral crypts and Polynesian feather capes to the diaries of Peace Corps volunteers.

In some respects the study of man here seems to know no bounds, but in fact the work concentrates on the physical and the previous. For the fraternity of physical anthropologists, "making some sense out of people" involves a kind of reflective counterpoint in which knowledge of ancient

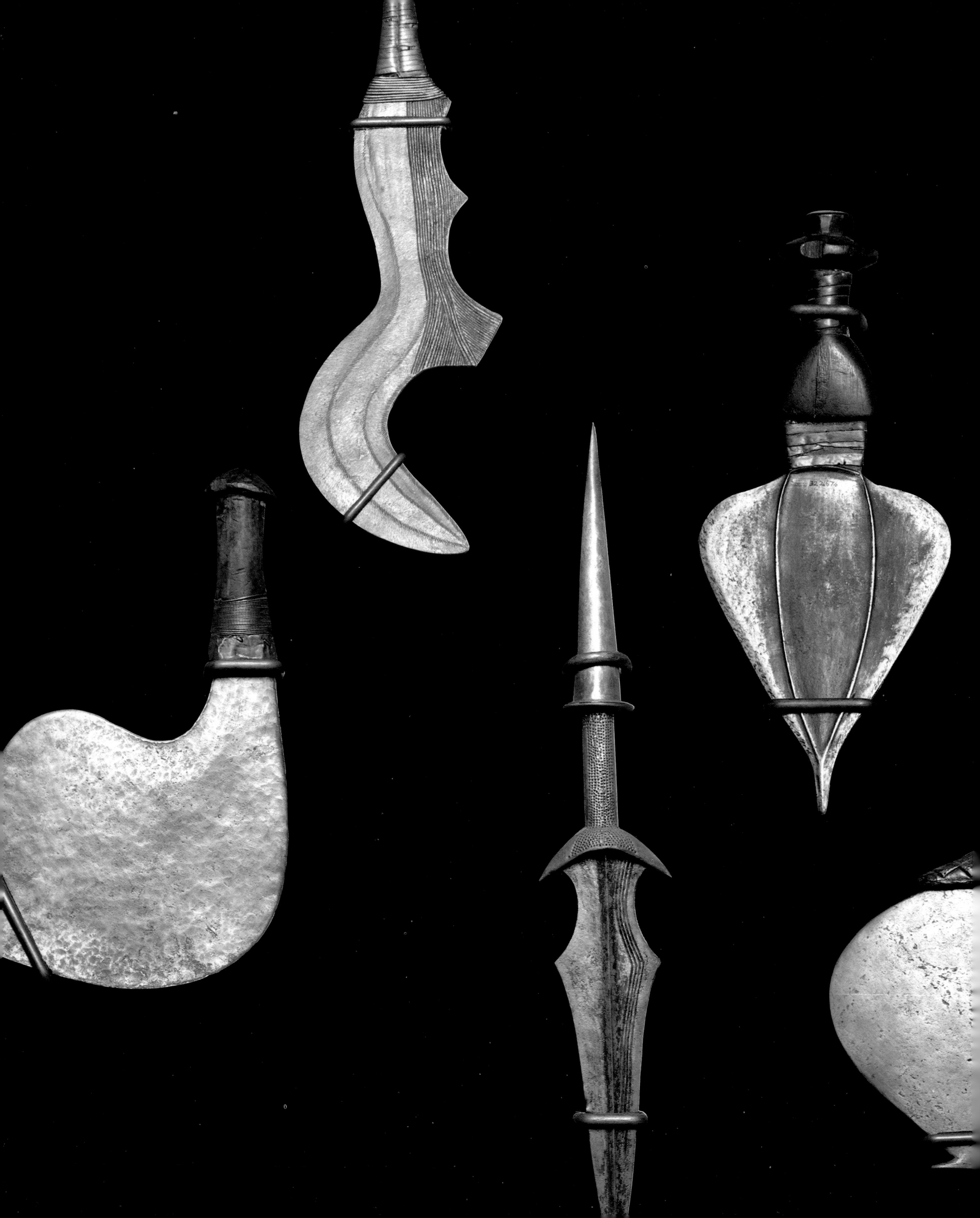

Opposite: *Contents of the Regourdou Cave in southern France, which inspired this diorama, suggest that Neandertal man practiced elaborate death rituals 70,000 years ago. The cave contained brown bear skeletons and the ceremonially bound body of a young man placed on a bearskin amid offerings of meat, stone tools and deer antlers. Nearby more bear remains were covered with a stone slab weighing a ton. The first anthropologist to describe Neandertals worked with a skeleton deformed by arthritis and showed him as a slouching brute. Later research reveals that these people, members of our own species, were very much like us. They even had a religion, commonly a prerequisite for culture.*

man illuminates his posterity, that is, of course, ourselves. Indeed an old motto might be inscribed over these laboratory doors: *Hic locus est ubi mortui vivos docent*—"this place is where the dead teach the living." The surprise lies in the variety of lessons that a single bone or collection of skeletons can reveal. An undisturbed burial ground, properly unearthed, discloses a population profile of the forgotten people who kept it: the ratio of men to women, the diseases they suffered, their stature, diet, nutritional standard, longevity and fertility. (J. Lawrence Angel has gone so far as to correlate muscle scars on pelvic bones with the number of children a woman bore.)

Surprising connections regularly arise among diverse cultures, disciplines and times. For example, discoveries in prehistoric Ecuador relate to an odd affliction among a literally small group of modern executives. Studying the skeletons found in ancient urns, Douglas H. Ubelaker found a consistent orthopedic anomaly in certain foot bones: a series of bumps on the metatarsals where the toes attach. Among modern men, x-ray data suggest similar growths among only one "population" that Ubelaker could locate: short people who work in offices. What could these folks have in common with the ancient Ecuadorian Indians? Both flex their toes backward for long periods, causing toe bones to rub against foot bones and raise calcified swellings similar to calluses. The contemporary people who display this benign syndrome have spent their professional lives in too-tall swivel chairs and gotten a purchase on the floor with acutely bent tiptoes. Ecuadorian Indians of every stature habitually performed everyday domestic tasks on their knees with their toes bent under them. Their artifacts suggested this, and the skeletal remains confirmed the hypothesis.

Cemeteries contain vast lodes of organic and cultural information. Donald J. Ortner excavated Bronze Age catacombs in Bab edh-Dhra—the likely site of Sodom—to learn that the people who lived there suffered from tuberculosis, among other things that afflict us still. (Physical anthropology has uses in the immediately contemporary world as well as the ancient one. The F. B. I. and police departments around the country often send unidentified human remains to the Museum, notably bodies found in the wild after months or years of exposure. The curators often close the books

The hunchbacked figurine comes from Jalisco, Mexico. This deformity was commonly depicted in Pre-Columbian art, suggesting that it was considered a sign of supernatural favor.

Opposite: *A Pre-Columbian stone jaguar head from Costa Rica.*

on routine "missing persons" cases; sometimes they provide key evidence in murder trials.) Less pathologically, ancient remains reveal how different societies interacted, the influences one had on another, the ways cultural habits changed.

For example, in their early-nineteenth-century descriptions of the Arikara Indians, Lewis and Clark said this South Dakota tribe buried their dead in individual graves soon after death—unlike neighboring tribes who placed their late relatives on elevated litters. Recent Arikara burial sites confirmed the explorers' reports but slightly earlier sites showed signs of secondary burial; skeletons were incomplete or lay in unlikely positions. Still earlier sites revealed something else again: the soil contained molted exoskeletons of fly pupae (maggots), suggesting that the bodies were exposed to open air for some weeks. The soil of even earlier burial grounds held the remains of different and very fastidious insects, certain beetles known to consume flesh only after it has become thoroughly dried out. Thus the Arikara at one time practiced open-air funeral rites like those that survived among their neighbors; then the Arikara adapted the practices of French explorers and missionaries. This finding has implications for extrapolating the relative adaptability of these people to alien visitors and also suggests that they were initially more like their neighbors than later observers had supposed.

For New World archaeologists, the phenomenon of interaction offers one of the great puzzles in human history, one complicated by the complex and still only dimly understood waves of prehistoric immigration to these continents from Asia. One anthropological consensus holds that the Eskimos, arriving 8,000 years ago or so, slowly spread across the barely habitable tundra that no earlier immigrants had the fortitude to settle. Curator William Fitzhugh spent more than a decade exploring the coast of southern Labrador to stimulate new thinking about the whole sequence. Along this serpentine coast of fjords, marshes, beaches and barrier islands, he discovered traces of 1,000 settlements. To everyone's surprise the oldest and most stable were Indian hamlets founded as early as 8,000 years ago. Furthermore, these "Maritime Archaic" Indians possessed a complex culture, a fact demonstrated by the sophistication and variety of artifacts that remain. One site, evidently a seasonal settlement, was occupied every summer for a few consecutive centuries by a community numbering several hundred people. While it was previously believed that the Indians clung to the forests as they followed the retreating glaciers northward, Fitzhugh showed that they left the woodlands entirely and pushed several hundred miles farther north than anyone lives today. But still they constructed longhouses made of wood which they evidently imported from farther south. When Eskimos finally arrived in the area about 4,000 years ago, they borrowed harpoon technology from their better-established neighbors. Both groups hunted birds, fish, walrus, seals and whales; both also mined a local quarry for chert. This stone, made into weapons and tools, was traded down the coast as far south as Maine.

For another New World investigator, "making some sense out of people" involves the unusual scientific exercise of hacking an elephant carcass apart with bone tools. Dennis Stanford was already skilled in the arcane arts of making flint arrowheads and using spear-throwers when he came upon a peculiar "kill site" that a bulldozer operator uncovered in suburban Denver, Colorado. It contained the remains of several mammoths, those pachyderms that became extinct in North America before people ranged the continent—according to conventional archaeological wisdom. But too many of the bone flakes seemed unusual and like one

Opposite: *This pottery house model bears witness to a fascinating event. The style of the seated figure, which conforms with a wealth of Pre-Columbian material, proves its South American origin more than 1,500 years ago. The architectural detail and the concept of a model house link it to Southeast Asia, specifically Indochina's expansive Han period. Acquired in coastal Ecuador before amateur collecting inspired counterfeits, the object suggests Oriental contact with the New World early in the Christian Era.*

Opposite: *Europeans brought horses to Argentina in the eighteenth century and nomads of the grasslands adopted them quickly. Here a Tehuelche hunter aims his bola at a flightless rhea, South America's version of the ostrich.*

another to Stanford's experienced eye. They were not the result of natural fractures or the breakage that occurs during predation or when carrion animals disturb a corpse. They looked intentionally wrought. Also, the archaeologist spotted among the bones a 30-pound boulder which, from a geological standpoint, probably shouldn't have been there. One likely explanation: someone put it there when mammoths ranged the Rockies and used it as an anvil to break bones. That way, hypothetically, toolmakers would have acquired the wherewithal for new spearheads, scrapers and cutting tools, which they could then use to kill and butcher more mammoths. But this all ran against the grain of the accepted chronology that held that the mammoths died out before people came along.

Stanford tested each element in his hypothesis. Using some of the bone flakes in thong-bound wooden handles which he fashioned, he determined that a man could carve up something as big as a mammoth—in this case, actually an elephant named Ginsberg that died in a zoo—and turn its useless bulk into manageable pieces of food. He also found that with stones used as hammer and anvil he could make flakes from elephant bones like the ones he'd discovered in the ancient site. Further, one series of experiments failed: the series designed to cause the odd bone fractures under purely natural conditions, that is, without human intervention. Large animal bones were fed to animals in the National Zoo; he placed them in other elephants' captive habitats, where the living animals toted them about, trampled on them and used them like clubs. Under naturalistic circumstances they didn't break into the characteristic flakes.

The reason for all this peculiar work: the oddly broken bones and presumed anvil stone could be 3,000 years older than the oldest previously confirmed human artifacts on this continent. Stanford's educated guess is that some previously unidentified people lived in the Denver environs 10,000 years before any other known Americans—before the Eskimos, before the previous wave of Na-Dene people (Athabaskans, Navajos, Apaches and others of the northern forests) and before "everybody else" currently known. If this Colorado site proves that human occupancy occurred much earlier, the prehistory of man in America will soon have to be rewritten.

Modern textiles, pots fired in antiquity and human skeletal material of any age—the materials of ethnology, archaeology and physical anthropology respectively—all find counter space in the Anthropology Processing Laboratory. All objects entering and leaving the collection storage areas must pass through the Processing Lab, which is responsible for seeing that all specimens are properly catalogued, recorded, handled, and stored.

For an Old World archaeologist whose life work centered on the Near East, "making some sense out of people" means writing the biography of an ancient town that arose in the land of Canaan more than 1,700 years before Christ. The early Canaanites of Tell Jemmeh built a community on a small plateau rising 45 feet above the arid plain about 10 miles from the modern town of Gaza. For 1,500 years it was variously a military outpost on the frontier, a way station for caravans bringing frankincense and myrrh from southern Arabia a two-month trek away, possibly a camel farm, finally a granary. In sum, generations of people sustained themselves much as the inhabitants of any border town in modern times. While they depended in part on their own local agriculture, they also profited from international commerce (and no doubt suffered from international conflict).

The tell had been dug before, by a noted Englishman who observed the rough-and-tumble archaeological conventions of his day: employ a gang of local laborors to dig wide swaths, unearth intrinsically valuable

artifacts and crate them. Then move on. This selective and brutal approach barely skimmed the cream of ancient sites but left them in shambles. The modern approach demonstrates adroit advances that archaeology has made. Scientists studying the past now recognize they cannot know what sorts of analytical feats their successors may perform in the future. "We can't even say what questions they'll ask a generation from now," says Gus Van Beek. Excavating a site inevitably destroys it in some respects. Thus the modern investigator preserves as much as he can by documenting every aspect of his work: recording the precise location of each object as it's found, whether it can be immediately identified or not; preserving everything but the dirt itself, and even saving samples of that after sifting it through fine sieves.

In former times archaeologists saved only objects of obvious value. In Egypt, for example, where archaeology began, the tombs of the Pharaohs were excavated and their carefully wrought contents hustled off to museums

Overleaf: *The Museum's Thai collection, renowned for its depth and variety, attracts scholars from around the world. The panoply of masks and bowls assembled for this photograph includes recent acquisitions, material bound for loan to other study centers and objects chosen for special research.*

27,156
SIAM
27,153
SIAM

and private collections. Even an old cemetery of commoners' graves was infamously destroyed save for two skulls because it held non-royal remains. This represented an incalculable loss of skeletal material to modern anthropology. As for such things as animal bones, to the horror of their successors, the Victorians considered them trash. Witness Tell Jemmeh again.

In cautiously digging down through 45 feet of earth and debris to virgin soil, Van Beek discovered traces of continuous human habitation between 1700 and 200 B.C. Among these traces were bones that zoological colleagues identified. Nearly 500 of them belonged to camels, an enormous number in comparison with a similar dig at Beersheba which uncovered only about 20 camel bones. Explaining the presence of unexpected objects like these represents a relatively new task: to resolve the inherent conundrums presented by whatever turns up, not just to look for old bronzes and pretty pots.

The concentration of camel bones may suggest a livestock farm or market. It also may be derived from camels that died while serving in caravans passing this way en route to Gaza, a Mediterranean port. They brought the riches of ancient southern Arabia: the resinous treasures that the Magi gave the Christ Child, according to St. Matthew's account. Frankincense and myrrh, both highly valued in the ancient world, were variously used in death rites, to scent temples and palaces, in making perfumes, to cure ailments from hemlock poisoning to menstrual problems.

The paleozoologists at Tell Jemmeh discovered other animals as well. Many were wild, though it appears that these were most common early after the tell's human settlement; by 1200 B.C. they vanished altogether, the likely result of permanent human settlement and ancient urbanization. Less easily explained: the remains of a giant African ostrich and a hartebeest, two exotic animals that were indigenous to the distant interior of Africa. How to explain their presence? Just as Mao Tse Tung gave President Nixon a pair of Chinese pandas for the National Zoo, perhaps ancient dignitaries presented zoological oddities to each other.

Oddities aside, the more substantial fruit of the new archaeology results from the most painstaking aspects of the Tell Jemmeh project.

Opposite: *A Haida mask from British Columbia has copper teeth and rolling eyes.*

Opposite: *Carved human figures and inscrutable ideograms adorn the foot-tall ceramic vase, a prehistoric Mayan artifact from Salvador.*

The site was covered by large silos in the third century B.C. which were apparently destroyed by an earth tremor. One, excavated by Van Beek, was equipped with hundreds of earthenware jars; they shattered. Van Beek plotted every sliver and chip of pottery as he removed them layer by layer, then shipped them back to the Museum. There, in a large basement room, a team of volunteers—perhaps 150 in all during the course of a decade—have pieced together scores of pots, jugs and amphorae. This painstaking work requires extraordinary patience and rather special talents: the ability to remember and compare small shapes, thicknesses, arcs and textures. An able worker might piece together two or three sherds on a good day and none at all on a slow one.

The purpose of this work is to compile a pottery chronology, a dating guide that can be used elsewhere through the Near East. Pottery styles changed slowly over time. A specific shape, the ornament on a lip, the curve of a handle or turn of a base serves as a reliable indicator of its origin in time. Each kind of pot, once dated in a dig as carefully researched as Tell Jemmeh, then offers a precise dating index.

"Making some sense out of people" means seeing each culture clearly—without the filters of one's own cultural bias. Americans tend to regard the maritime traditions of Oceania as inexplicably epic. How could the Polynesians, we ask, have ventured across 2,000 miles of open ocean to the most remote islands of the world? After all, the first of them couldn't have known the islands were there. From the Polynesian perspective, however, it was simple. They'd inhabited islands for the many generations since the Chinese hegemony forced their ancestors to leave the Asian mainland. Islands were their home; the sea was full of them and there were always more so far as they knew. It was simply a question of setting forth in seakindly boats, living off the bountiful sea and the things they carried with them, then finding the next island—a prospect that can be likened to sodbusters trekking along the Oregon Trail with wagons and wheelbarrows.

"Making some sense" also means appreciating the many differences among people, their views of the world and their various places in it. These tenets can differ to the point of diametric opposition. In West Africa, the Bantu, the inhabitants of a flat expanse, have special regard for the

This contemporary dance mask from Bali depicts the traditional witch, Rangda. Using her "black" magic, she entrances dancers armed with kris knives so that they turn the curved daggers on themselves—until Barong uses his "white" magic to defeat her.

Opposite: *The turn-of-the-century entertaining Mexican dance mask comes from a rare set of twelve. Made for clowning at fiestas on Christian holidays, it preserves (according to Donald Cordry) at least one element of Pre-Columbian culture: the curly beard worn by the old rain god.*

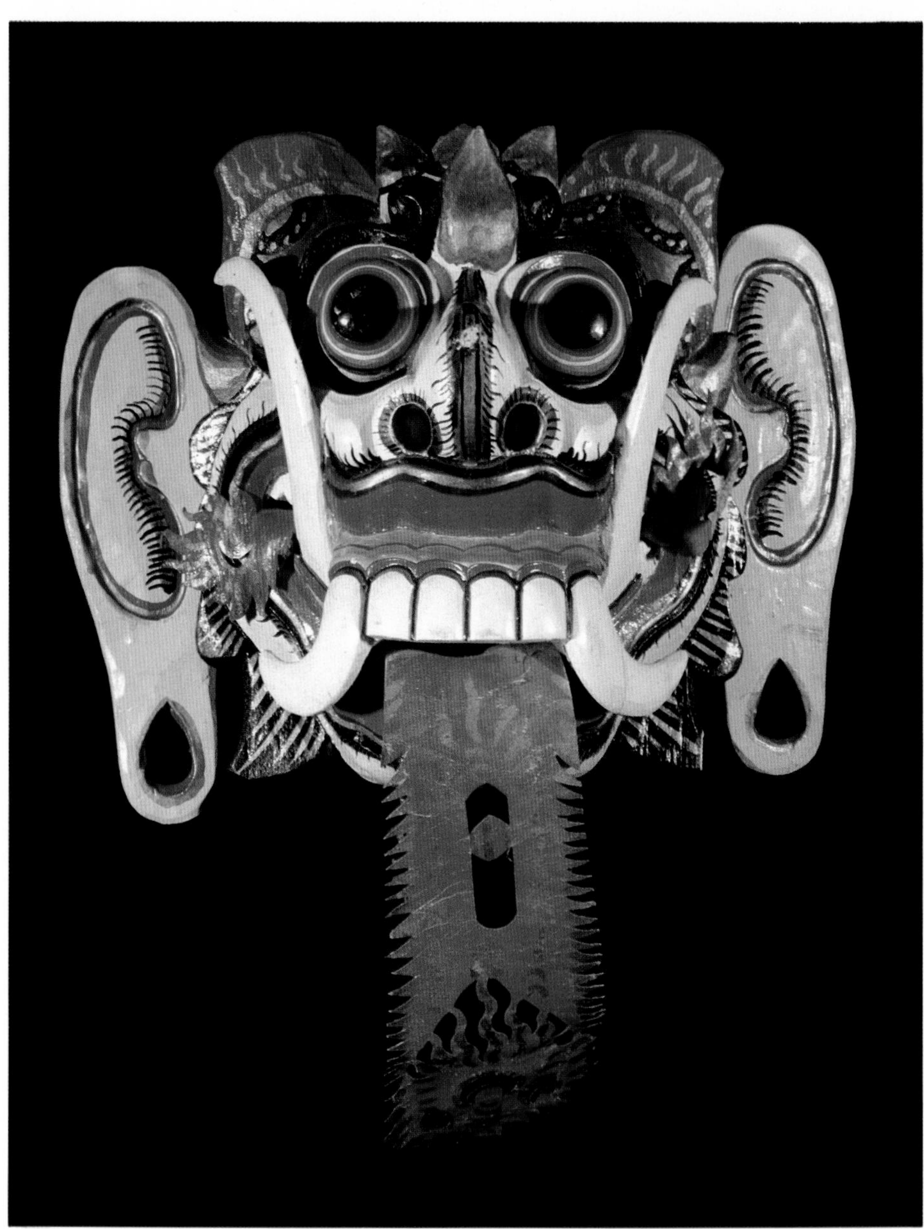

single hill that rises above the level plain. One of their myths relates that it was on this hill that the first man descended from the sky to earth on a rope. When he alit, the rope was hauled back into heaven. Contrarily, a third of the way around the world, the Hopi Indians believed that their ancestors inhabited the bowels of the earth and emerged up into it through a hollow reed. A third of the world away again, New Guinea pygmies tell of a patriarch who climbed down a rope to go hunting and fishing and was stranded—similar to the first Bantu. His wife tossed fire down to him, as Prometheus brought fire to the Greeks. Some traditions hold that man emerged from dust; others say in the beginning was the slime. Adrienne

Kaeppler, an expert in oceanic ethnology, quotes one version of Hawaiian lore:

The slime, this was the source of the earth
The source of the darkness that made darkness
The source of the night that made night
The intense darkness, the deep darkness
Darkness of the sun, darkness of the night
  Nothing but night.

The night gave birth
Born was Kumulipo in the night, a male
Born was Poéle in the night, a female
Born was a coral polyp, born was the coral, came forth . . .

Then a grub and its child, an earthworm, then the starfish, the sea cucumber, the sea urchin and short-spiked sea urchin and smooth urchin and ring-shaped urchin, the barnacle and his child the pearl oyster, the mother of pearl and the oyster, the mussel, the hermit crab, the limpet, cowry, conch, the nerita, the sand-dwelling snail, the freshwater snail. Then

Born was man for the narrow stream,
The woman for the broad stream . . .

Among all these people and myths, one commonality appears: the propensity of people to explain their own origins, and in so doing to ascribe a special kinship with the forces of creation. By studying people, in another way the Museum of Man pursues the same goal enlarged: to define manifold mankind, explaining our place and past within earth's context as we strive best to understand it—within natural history.

Opposite: *Urns like this, possibly filled with food, incense or water, were interred in upper-class tombs of Mexico's Zapotec culture. The central face may portray the deceased, the birds under the arms eagles. The headdress figure is a mythical exotic, somewhat like the later feathered serpent Quetzalcoatl, whom the Aztecs revered and for whom the Dinosaur Hall's pterosaur is named. Like the Museum itself, what a remarkable nexus.*

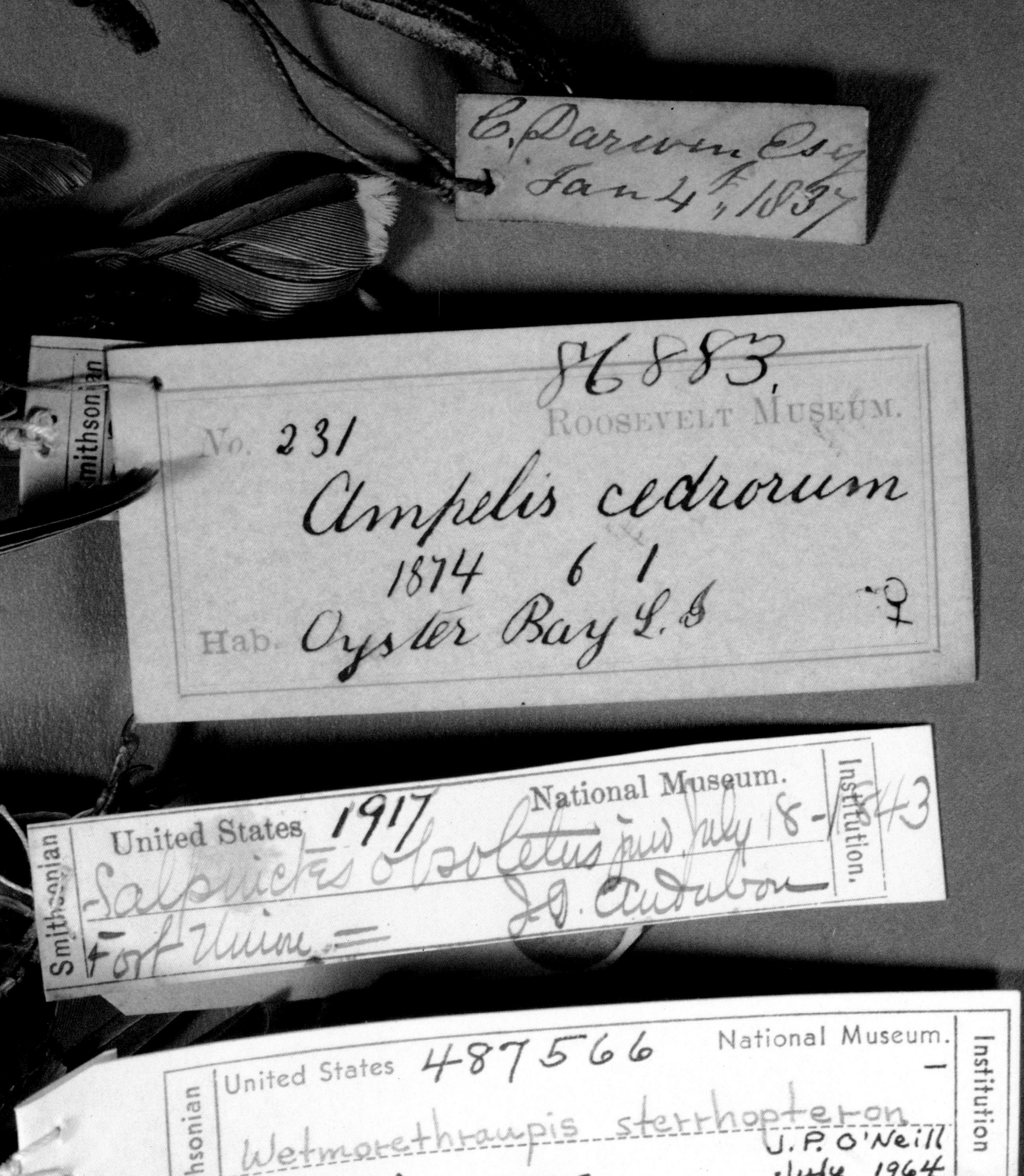

C. Darwin Esq
Jan 4th, 1837
86883
Roosevelt Museum.
No. 231
Ampelis cedrorum
1874 6 1
Hab. Oyster Bay L. I
Smithsonian
United States 1917 National Museum.
Salpinctes obsoletus juv. July 18-1843
J. J. Audubon
Fort Union
Smithsonian Institution.
United States 487566 National Museum.
Wetmorethraupis sterrhopteron
J. P. O'Neill
July, 1964
PERU: Amazonas
Smithsonian Institution

INDEX

Numbers in roman type refer to text pages. Numbers in *italic* type refer to pages on which illustrations appear and also refer to information included in captions for that illustration.

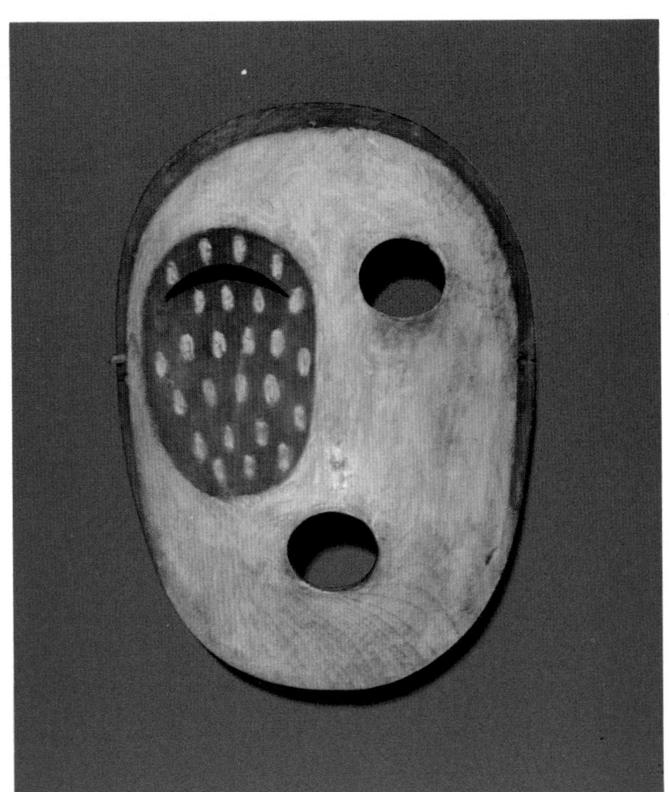

CREDITS

The numbers refer to pages on which the photographer's work appears.
t = top, b = bottom, m = middle

CHIP CLARK
2–3, 4–5, 10–11, 14, 18, 24, 27, 28–29, 30–31, 32, 55, 56, 57, 58–59, 61, 62–63, 66, 71, 78–79, 80b, 84–85, 86, 87, 88–89, 91b, 92, 93, 94, 95, 96, 97, 98, 102–3, 106, 114–15, 116–17, 119, 120–21, 122–23, 128–29, 130, 131, 132, 133 both, 134, 135, 136, 155, 160–61, 164, 165 all, 172–73, 184–85, 189, 190, 196–97, 198–99, 200 both, 202, 203t, 204–5, 206–7, 209, 210–11, 212–13, 214–15, 216–17, 219, 220, 221, 222–23, 224, 225, 226–27 all, 228–29, 230, 231, 232–33, 234–35, 237, 238, 239, 240–41, 242, 243, 244b, 246, 256–57, 258–59, 260, 262–63, 264–65, 267, 274, 276–77, 278, 281, 282–83, 284–85, 286, 287, 288–89, 290–91, 293, 296, 297, 298, 302, 304–5, 309, 310, 312, 313, 314–15, 316–17, 318, 319, 320, 321, 323, 324, 325, 326–27, 328, 330, 331 both, 332, 334–35, 336–37, 344–45, 347, 348, 349, 350, 351, 360, 361, 362, 366–67, 369, 376–77, 379, 394, 395, 412, 414–15, 417, 418–19, 420–21, 423, 425, 426, 427, 428–29, 437, 438, 440–41, 443, 455, 465, 466–67, 468–69, 471, 477, 478–79.

KJELL SANDVED
6–7, 8–9, 12–13, 20, 26 both, 52, 60, 69, 72, 73, 74–75, 76, 80t, 81, 83, 90, 91t, 100 both, 101, 104, 105, 108, 109, 110–11, 113, 118, 124–25, 127, 138–39, 141, 142, 143, 144–45, 146–47, 148, 149, 150, 152, 156, 157, 158, 159, 163, 166, 167, 168, 170, 174 both, 177, 178, 179, 180, 181, 182, 183, 186, 187, 188, 192–93, 195, 201, 203b, 244t & m, 248, 250–51, 266, 294, 306, 339, 341, 342 both, 346, 352, 353, 354 all, 357 both, 358, 365, 370–71, 374, 381, 382–83, 384–85, 387, 388–89, 391, 392–93, 397, 399, 400, 403, 405, 407, 430, 432–33, 434–35, 445, 446–47, 449, 452, 456, 457 both, 458–59, 460–61, 462–63, 474, 480–81, 482, 485, 486, 487, 488, 490, 496.

The gems seen on page 56 were donated to the Museum by the following: G6186, Mr. and Mrs. D. C. Harrold; G8127, Mr. K.M. Kupper; G8884, Dr. and Mrs. R. L. Bonsanti; G8965, Dr. J.P. Gills; G7466, Dr. G. Hays; G9140, Dr. P. Gutlohn; G8975, Dr. J.P. Gills; G8932, Mr. C. Entemann.